PENGUIN BOOKS

SECRETS FROM AN INVENTOR'S NOTEBOOK

Maurice Kanbar and his inventions have been featured in articles in *The Wall Street Journal, Forbes, Business Week, Newsweek, USA Today*, and many more. His enthusiastic support of the arts, especially independent film, led to his endowments of NYU's film school, the Maurice Kanbar Institute of Film and Television.

SECRETS FROM AN INVENTOR'S NOTEBOOK

MAURICE KANBAR

With a Foreword by Phil Baechler

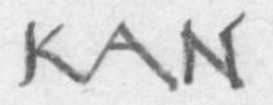

PENGUIN BOOKS

While we have tried to make the information in this book useful and accurate, much of it is based on the author's own experiences, and due to changing laws and the particular circumstances of any individual idea or product, the author and the publisher cannot assume responsibility for any action readers take based upon material offered in this book. For legal advice or other expert assistance tailored to the specifics of your idea and situation, seek the services of a competent professional.

PENGUIN COMPASS
Published by the Penguin Group
Penguin Group (USA) Inc., 375 Hudson Street, New York, New York 10014, U.S.A.
Penguin Books Ltd, 80 Strand, London WC2R 0RL, England
Penguin Books Australia Ltd, 250 Camberwell Road, Camberwell, Victoria 3124, Australia
Penguin Books Canada Ltd, 10 Alcorn Avenue, Toronto, Ontario, Canada M4V 3B2
Penguin Books India (P) Ltd, 11 Community Centre, Panchsheel Park, New Delhi – 110 017, India
Penguin Group (NZ), cnr Airborne and Rosedale Roads, Albany, Auckland 1310, New Zealand
Penguin Books (South Africa) (Pty) Ltd, 24 Sturdee Avenue,
Rosebank, Johannesburg 2196, South Africa

Penguin Books Ltd, Registered Offices: 80 Strand, London WC2R 0RL, England

First published in the United States of America by Council Oak Books, LLC 2001
Published in Penguin Books 2002

5 7 9 10 8 6

THE LIBRARY OF CONGRESS HAS CATALOGED
THE HARDCOVER EDITION AS FOLLOWS:
Kanbar, Maurice.
Secrets from an inventor's notebook / Maurice Kanbar.—1st ed.
p. cm.
Includes bibliographical references and index.
ISBN 1-57178-099-8 (hc.)
ISBN 0 14 20.0056 6 (pbk.)
1. Inventions—Marketing. I. Title.
T339.K23 2001 608'.068'8—dc21

Printed in the United States of America
Set in Optima / Designed by Melanie Haage

contents

Foreword by Phil Baechler vii

Preface xi

Introduction: *The D-Fuzz-It Story* 1

The Five Steps to Inventing a Success 7

1 Solve a Problem

- ► You Oughta Be in Pictures: The Quad Cinema 9
- ► What's Wrong with This Picture? Observation and Curiosity 16
- *Problems: Yours, Mine and Ours* 21
- ► The Real World: Move from Idea to Invention 25
- *Reality Check: Reinventing the Wheel and Other Common Mistakes* 33

2 Prove Your Invention/Build a Prototype

- ► Ouch! The Kanbar Target and the ROLLOcane 37
- ► Two Popsicle Sticks and a Coat Hanger: Prove Your Invention Cheaply and Simply 43
- *Money, Money, Money, Money, MONEY!* 49
- ► People Watching: Observe the End User 56
- *Play by the Rules* 64

3 Protect Your Idea

- ► Better Safe Than Sorry: The Needle Protector 71
- ► This Won't Hurt a Bit: Understanding Patents 76
- *Defining Our Terms* 78
- ► Get It in Writing: Protecting Yourself 82
- *Conducting a Patent Search* 85
- *Applying for a Patent* 88
- *Hiring a Patent Attorney* 93

4 Manufacture or License?

- It's a Puzzle: Tangoes 99
- Making It: When to Manufacture 105
  - *Hiring a Manufacturer* 109
  - *How to Make It Even Better* 112
- Making a Deal: When to License 114
  - *What's the Deal?* 119

5 Market with a Twist

- You Go to My Head: SKYY Vodka 125
- First Impressions: Name, Package, Price 135
  - *Perfecting the Package* 141
  - *Making Sure the Price Is Right* 149
- What's the Buzz? Spread-the-Word Distribution and Marketing 152
  - *Publicity, Promotion and Advertising* 157

Afterword: The Fundamental Things Still Apply 167

Appendix

- Creativity, Inventing and Brainstorming 169
- Inventor's Reading List Addresses 170
- Prototyping, Fund-raising and Business Plans 170
- Documentation, Disclosure and Patenting 173
- Manufacturing and Licensing 177
- Marketing, Naming and Packaging 177
- Inventor Resources, Associations and Assistance 178

Bibliography 181

Acknowledgments 185

Index 187

foreword

I wish I'd had this book when I first invented the Baby Jogger! It's a long road from a wild idea to a commercially successful product, but Maurice Kanbar has written a delightful map to help navigate that route successfully: from prototyping and market research, to patenting, manufacturing and selling a finished product.

The first step after that lightbulb flashes on over your head is prototyping. For me, that consisted of welding some bicycle wheels onto an old rusty stroller I picked up for five bucks at a secondhand store. (Hey, don't laugh: some software billionaires started with Commodore computers.) In the chapter on prototypes, this would fall into the category of "functional prototypes." It was an ugly stroller, but it worked, and it was the first step in the evolution of the Baby Jogger.

One of the biggest bumps on the road to success can be protecting your patent rights. When I started almost two decades ago with little more than a handmade stroller, one of the few information resources I could find was a skinny brochure from the U.S. Patent Office. Trust me, what you're holding in your hand goes way beyond that meager information source.

Filing and enforcing a strong patent can be a complicated process, but Maurice's book is chock full of real world experiences and examples, both from his and other inventors' case histories, told in a tone that's closer to an action-adventure movie than a dry how-to manual. The patent office plays by strict rules, however, so even such procedural details as how to document the inventive process are clearly explained (When I start a new stroller design, I keep all the documentation in a "Baby Book!"). Most of us wouldn't do surgery on ourselves, so when it's time to find a good attorney, Maurice even has tips on selecting a firm to help you get the most bulletproof patent possible.

Of course, not all inventions are big winners. The old saying about "invent a better mousetrap and the world will beat a path to your door"—I can't remember the last time I bought a mousetrap. That's just the point! It's the NEW stuff that people are buying. I grew up listening to 45-rpm records. My kids don't even know what those are. The pace of innovation is increasing and there are always new gizmos coming on the market.

You may have such a gizmo, or just an idea. Many times people have heard I was an inventor and confided in me: "Hey, I've got an idea." Some of those brainstorms actually were pretty good, but most of the time I never see the person again, or their product. Let's face it, getting to market can be tough. Sometimes it takes a lot of cash, and I typically warn people not to sell their house to start a business. (When I started my business I didn't even own a house to sell!) It's possible you have an idea that's good enough to patent and then license to an established company. That route could have some pitfalls, though, and there are plenty of sharks out there who would love to "help" you get your invention to market, namely their market. There are even firms, Maurice points out, that may lure you into a license deal just to keep your product from competing with theirs. Whether you want to find a manufacturer to do the work for you, or simply turn the product over to someone else on a royalty basis, Maurice's detailed analysis of the process could help you avoid nasty surprises and get the best deal.

The bottom line, as they say in business, is that the consumers vote with their dollars. Selling a product is one of the mainstays of American life. Turn on your TV and you'll see what I mean. We are bombarded by messages, and getting attention in a crowded marketplace is a major challenge. It was a long road from the time I started running in races with the Baby Jogger, to the point at which it was showing up in movies or in tabloids being pushed by sunglass-wearing movie stars.

With SKYY Vodka, Maurice faced the same marketing challenges. How could a little upstart vodka in San Francisco take on the established brands? It took a lot more than just a blue bottle, so the book's final segment on launching a successful brand reads like a fast-paced MBA in marketing that even non-inventors can enjoy. Can I mention Jack Nicholson and David Letterman? Just did.

I don't know how many folks have called me for advice over the years and I've spent hours talking about prototypes, patents and marketing. Now I can tell them to get this book, because it's all in here. I especially liked the references listed in the Appendix: Yellow Pages for inventors! Much of my patent and litigation experience has been by the school of hard knocks—this book provides a much easier, more effective and cheaper way to quickly get on the right track. Not only that, it's a lot more fun to read than patent abstracts or legal briefs. So, if you have an invention or even just a glimmer of an idea, get your reading glasses and thinking cap ready!

Phil Baechler

preface

I've been inventing and launching new products since my early twenties. People have sometimes characterized me as a born inventor, and I suppose I wouldn't disagree—because inventing solutions to problems, improving devices and systems, and then taking those solutions and improvements into the marketplace has never felt anything but natural to me. For better or worse, I'm hard-wired that way. I'm grateful to have been able to turn my natural inclinations—constant curiosity about the world around me, a determination to learn how things work and then make them work better, and an interest in how and why we buy things—into a successful career.

As a Brooklyn born teenager, I spent several summers working in a summer camp in upstate New York. One summer I was in charge of the dining room, which was staffed by campers. I revamped the way we worked and created new systems and schedules. As a result the room ran smoothly, with none of the screw-ups or temper tantrums of previous summers. At the end of the season, my boss called me into his office. I was expecting a pat on the back, but instead I learned that he attributed our efficiency not to my innovations but to the exceptional group of workers we had that summer. I was disappointed but I learned a valuable lesson: innovate not because you want your efforts noticed but because *you* will know you've done your best. (And if at all possible, be your own boss!)

About fifteen years ago, I moved from New York to California. I'd had enough success with my inventions to kick back a little and enjoy a relaxed pace of life in San Francisco. But instead of slowing down, I stumbled on another problem I wanted to solve. In the process I invented and launched my most successful enterprise yet: SKYY Vodka.

Since my lesson in the summer camp dining room, I'd been content to do my work in the background, getting satisfaction from my successes and learning from my missteps without much outside attention. After SKYY's success, though, I began to find that people were interested in how I did what I do. Business and media people were curious about how someone could come out of nowhere and be so successful in the spirits business. They wanted to know how an inventor could launch a *vodka*. Home-based tinkerers and inventors wanted to know if I could teach them how to develop and profit from their innovations. Instead of routinely being asked, "Are you crazy?" (a question inventors soon get used to), I started being asked if I could help other people learn how to be similarly "crazy." And publishing types started asking me to write a book.

Some of those publishing types were from Council Oak Books. Before anyone cries foul, full disclosure: Several years ago I purchased Council Oak Books, a small, respected but struggling publishing house. As you'll see in the following chapters, I don't relish the day-to-day running of businesses, and though I'm an avid reader, I had no interest in being a book publisher. So I hired talented and experienced book people to maintain and improve Council Oak. Their long efforts to pry a book out of me about my inventing life were finally successful for two reasons.

First, they reminded me that all author royalties from the book would be given to charity. One of the joys of making money is being able to give it away to worthy people and causes. When you have more than you need and realize you can't take it with you, you start giving it away.

Something I read long ago in a biography of Andrew Carnegie has stayed with me: He gave much of his enormous wealth away before he died and he noted that it had been much easier to make his fortune than to spend it intelligently. I'm certainly no Carnegie, but over the years I have taken great satisfaction in providing financial help to young filmmakers through The Kanbar School of Film and Television at NYU, and in supporting worthy causes like cancer research through the Strang Center and the ecological efforts of the Wildlife Conservation Society. All the author proceeds from the sale of this book will be donated to charitable organizations and enterprises through the Kanbar Foundation.

> The value of a dollar is social, as it is created by society.
>
> — *Ralph Waldo Emerson*

Second, after agreeing to give a talk one evening at The Learning Annex, I was persuaded that hearing about my experiences might be useful and even inspirational to aspiring inventors and entrepreneurs. That class was packed with people of all kinds who listened intently and couldn't ask enough questions. There are a lot of curious, thinking people out there with great ideas, and they're keen to hear how others have made ideas pay. I know I didn't have a lot of mentoring or encouragement as I was coming up. Though I think I was born to invent and don't think anyone could have stopped me from persevering until I'd done it successfully, a little friendly advice wouldn't have hurt. And as I looked back over my business life, I saw that I hadn't been simply following my gut; I'd been following some rules, rules that anyone could learn.

Plus, I truly revel in the successes of other inventors, some of whom you'll meet in this book. When a Sam Farber (OXO Good Grips kitchen utensils), or Phil Baechler (the Baby Jogger), or Ann Moore and Lucy Aukerman (the Snugli) parlays his or her hard work and ingenuity into a marketplace winner, I'm filled with admiration.

I hope reading about what has worked for me—and what hasn't—helps you. The basic how-to steps of inventing aren't mysterious. But executing each step and inventing a *success* can be confusing and difficult. If it weren't, every new product launched would be a hit. And most ventures, even in our current boom economy, are not.

I've written this book both to illustrate the concrete steps inventors take (from brainstorming and brand naming to prototyping and patenting) and to demystify the process so that you'll feel more confident about taking your first or next step. I look forward to reveling in your success.

Maurice Kanbar
San Francisco

introduction

THE D-FUZZ-IT STORY

The dude ranch was my friend Martin's idea. It was the mid-sixties and he'd heard this was a good place to meet women. So there we were, a couple of twentysomething Brooklynites, doing our best rugged cowboy impersonations.

I was leaning against a wall, trying to look cool. I don't think my red sweater vest helped, but I always wore one. Martin was chatting up a couple of likely candidates and motioned for me to join him. When I pulled away from the wall, I noticed some pills from my sweater had clung to the concrete. "What are you looking at, Maury?" whispered Martin. "Get over here!" But it was hopeless. I couldn't stop thinking about how efficiently the wall had removed the annoying balls that form on sweaters. As usual, I couldn't rest until I figured out how the wall had restored my sweater.

I was used to people like Martin getting impatient with the way my mind works. I'd been the kind of kid who not only asked why, but also how. It drove my parents nuts! My father would come home from a long day of work at the small laundry he owned and settle into his easy chair with the paper, only to have me ask how the lamp by his chair worked. He'd mumble something about electricity coming from the wall into the lightbulb, but that didn't cut it for me. I needed to know how electricity *worked.*

I studied the rough texture of the wall instead of trying to charm our new friends. It must be the sharp edges of the sand crystals in the concrete that pulled off the fuzz, I thought. Could I replicate the wall's texture in a small gadget I could use at home? I thought such a tool would be a practical way to care for sweaters, much better than the brushes that were currently on the market. I knew I'd buy one, so I got to work.

Though I'd never invented anything, I knew that an invention is a *thing,* not just an idea. I'd studied engineering and chemistry in college, so when I got home I began experimenting. I bought some coarse fabric and sprinkled it with glue and aluminum oxide crystals (the stuff in sandpaper). I figured that it was a bit like concrete, with its rough surface, and might grab. I let it dry and then ran it over one of my sweaters. It worked beautifully. My crude but functional model was simply the abrasive-coated fabric stapled to a piece of wood.

I wanted to protect my invention with a patent and I had heard that you don't need a clean and shiny working model in order to apply for one. The Patent Office stopped requiring them when they ran out of storage space! My college roommate's cousin Herbie was a patent attorney, so I went to see him for help. He knew this was to be my first patent and wanted to make sure I wasn't too starry-eyed about the whole thing. "Look," he said, "I want you to understand that probably one in ten applications for a patent is granted. But out of those issued patents, I'd say that only one out of a hundred is ever manufactured. What's more, of those that are manufactured, maybe one in fifty is successful. You'd better be confident." I was. You've got to have guts to stick with your invention because many successful ventures were once considered ridiculous by lots of people. Despite Herbie's cautions, I applied for and was issued a patent.

My next step was to create a better model, something closer to the actual product I hoped to sell. I envisioned an inexpensive, durable plastic handle. A few blocks from my home there was a shop with a "Plastic Molding" sign in the window. I went in and talked to a model maker who said that sure, he could make a mold and prototype of my device—for $1,200. Unfortunately, I had used up my savings on another scheme that hadn't yet come to fruition. While working for a company that distributed Dupont fabrics, I had devised a unique method for manufacturing nylon fibers at a price competitive with that of Dupont. It would take several tries, though, before the pilot operation clicked, and I was eventually able to sell the enterprise to a Fortune 500 company, the largest women's hosiery manufacturer in the country. But that happy ending was a few years away and I needed the cash *now* for my plastic handle.

I tried to get the $1,200 for the mold at a few banks, but none of them would lend me money because I had no money to begin with. They only like to provide loans when they *know* you can pay them back! So I asked my mother for it. "Don't be crazy!" she said. "Why do you want to do that when you have such a nice job? Keep your job, honey."

But I didn't want to keep my nice job. I'd long known that I didn't want to be someone else's employee. So I saved the $1,200 by pinching pennies for eight months. I even told my dates they could only order a hamburger and a coke when I took them out because I was on a strict budget. I stuck with my strict budget because I had faith in my idea—and I'm stubborn.

When I'd saved the money, I had a one-cavity steel production mold made. A four-cavity mold would have made four times as many pieces in the same amount of time, but it would have required a greater initial financial outlay. We made our first samples with the one-cavity mold and invested in a four-cavity when the number of orders warranted it. (Each additional cavity in a mold significantly increases its cost. Even if one can afford more, it's prudent to do initial production runs with one- or two-cavity molds.)

Mold in hand, I started working on packaging. Because I was broke again, I couldn't afford to pay a graphic designer the $2,000 he wanted to design a hanging card to hold the device. While I didn't know a thing about packaging, I did know that cosmetic companies spent a lot of time and money perfecting theirs. So I went to drugstores and studied make-up boxes and bottles of shampoo. I bought products that featured typefaces and designs I liked and wrote some simple, descriptive copy:

"Brushes away sweater fuzz in seconds . . ." I talked a friend who was between jobs into going in on the venture with me for a percentage of the business. Together we scraped together $50 to pay another designer to do up our copy and design. And *voilà,* we had a professional-looking product. I actually wanted to call my invention "Balls Off!" but this was 1964 and everyone said, "Are you crazy? [I get asked that a lot.] You can't call it Balls Off!" For once I relented and went with a name my friend Helen Shufro came up with: the D-Fuzz-It® Sweater and Fabric Comb.

The next problem—and it seemed like a big one—was how to get the D-Fuzz-It into stores. We certainly couldn't afford to hire a sales rep. I went to a library reference room and found a list of department stores, with contact names and addresses. I wrote a short letter to the buyers in sweater departments. At the bottom of the letter, beneath a dotted line and the words, "cut here," I included a simple order form. I mailed off hundreds of these, along with sample D-Fuzz-Its. Because the buyers could try the sample and see that it worked, I got orders. Hundreds of orders, then thousands.

Each D-Fuzz-It cost us about 14¢ to make. We sold them for about 40¢ and they retailed for 98¢. We made $200,000 that first year, quite a lot of money at the time. About six years later, I let my partner buy my part of the business. All these years later, the D-Fuzz-It sweater comb is *still* a reliable profit-maker. My package copy has even been translated into several languages because the comb is now sold internationally.

"I'm glad I was wrong," my mother told me at the

time. By my thirtieth birthday, I never needed to work for anyone else again. No more dude ranches either. Not long after, I found myself on the *David Susskind Show* as one of the Five Most Eligible Bachelors in New York.

THE FIVE STEPS TO INVENTING A SUCCESS

The Fundamental Things Apply

Many years, more than thirty patents, and several million dollars later, I still follow the same five fundamental steps I learned from inventing and marketing my first successful product, the D-Fuzz-It sweater comb.

1. Solve a Problem
2. Prove Your Invention/Build a Prototype
3. Protect Your Idea
4. Manufacture or License?
5. Market with a Twist

From the humble D-Fuzz-It to the high-end SKYY Vodka, nothing has changed my approach—not the technical Internet revolution nor the fact that banks now come knocking on *my* door. With experience, you get better at each of these steps, and you may sometimes be tempted to skip or scrimp on one or two. But there really are no shortcuts. Each step must be executed thoroughly and you must *ruthlessly* review your progress at every turn. Is this problem really worth solving? Do I have what it takes to solve it? Does my design meet each consumer requirement? Is this the perfect name? How can I make it even better?

Because these five steps are so important, the following chapters are organized around them. Each chapter begins with the story of one of my inventions, one that illustrates the task at hand. Then, because each step is made up of many others, we'll take a close look at the how, when and why of the real-world process. An appendix follows, filled with practical resources, suppliers, organizations, phone numbers, publications and Websites.

But let's start at the beginning, with a question inventors get asked all the time, "How did you get that idea?"

chapter one

SOLVE A PROBLEM

You Oughta Be in Pictures: The Quad Cinema

At a Manhattan dinner party in 1972, I met a young man whose family owned a number of movie theaters. He complained about how lousy business was. "TV is ruining us," he said. "People just don't go to the movies the way they used to."

As a regular moviegoer, I was surprised to hear this. I grew up when kids flocked to the theaters every Saturday afternoon to see the next episode of serials like *Flash Gordon* and *Buck Rogers*. On Friday and Saturday nights, when our parents made an evening out of dinner and a movie, there were lines around the block. Everyone seemed to go to the movies back then—and I still went.

But more and more, television was keeping families at home. Why go out when you could watch comedy, drama, variety, sports and game shows while eating a TV dinner off of a TV tray? Watching TV was easy and free, unlike hiring a baby-sitter, finding a parking space and buying tickets, popcorn and soda.

The movie theater fellow said that his 1,200-seat theaters were lucky to have 120 people in them on most nights. I knew that it was true that television offered lots of entertainment, and it made sense that young families would find it easier and more economical to stay in, but by his own admission, *some* people were still going to the movies. Who were they?

I resolved to find out and did some "market research," attending movies and counting the house. I saw that the theater owner was right: no matter what the size of the theater, there were generally only about a hundred people at a time buying tickets for any given screening. And the people I saw were mainly in their twenties and thirties. I concluded that young people were still going to the movies on dates. Why were *they* still venturing out? If this was the core audience that could be counted on, it seemed important to understand their motivation.

> Always search for the cause of something unexpected.
>
> —*Guy Kawasaki,* Rules for Revolutionaries

I often think problems through by concocting scenarios . . . and by talking to myself. Imagining real-life scenarios is a little like writing a screenplay, only in this case you're trying to understand what people are already thinking, and what's motivating their actions. Here is the situation I imagined for most of the people I saw going into the theater: Joe is interested in Mary and would like her to wind up at his place at the end of their evening. Joe could call Mary up and invite her over to see a movie on TV but she might think, "Not so fast, buddy" or "How cheap!" But if he calls her up and invites her out for dinner and a movie, he might just get her over to his place afterward. What better motivation could there be? There may have been a sexual revolution going on, but dating hadn't disappeared. And my common sense told me that there was also a percentage of the population that would always want the eventlike feel of going out to enjoy a film with other people. We crave the communal experience. Bottom line: Fewer people were interested in going to the movies regularly, but I concluded that there were some people who always would.

So, I reasoned to myself, TV had diminished the movie-going audience, and TV was here to stay, but there was still a smaller audience that could be relied upon. If you have a smaller audience, you obviously don't need a thousand-seat theater. How do you turn a profit when you are selling fewer tickets?

"What if you used the same space you'd use for a thousand-seat theater on several smaller theaters?" I asked myself. Rather than one movie bringing in a hundred people, you could have four movies each bringing in a hundred Joes and Marys. Instead of concentrating on increasing the number of customers for one film, why not increase the number of offerings? Without requiring any more ticket takers, concession-stand workers or projectionists, you could easily multiply your profit. And instead of complaining about your shrinking audience, you'd be giving that audience more viewing choices and a novel atmosphere.

I felt ready to build a model and test my idea. At the time, I owned two small buildings in Manhattan. I had a lab where I worked upstairs in one, but I couldn't rent the ground floors—hard to believe given today's crazy New York real estate market. I decided to turn the buildings into four small theaters.

I knew I needed a distinctive name, something modern that let people know this was a new kind of cinema. While the theaters were being built, I held a contest among my friends. "Come up with a catchy name," I said, "and I'll give you a year's pass to the movies." Soliciting ideas from a variety of people, of all ages and all walks of life, is a good naming technique. And contests are a tried-and-true method of involving potential customers in your process. The Planters Peanut Company got their Mr. Peanut graphic from a teenager who entered a contest they sponsored to create a trademark.

One friend suggested I call my theaters Movies 4. The name I liked best was the Quad Cinema. I gave the names my scenario test. When Joe called Mary, or given the seventies' sexual revolution, when Mary called Joe, would they say, "What's playing at the Movies 4?" or "Let's go to the Quad?" The Quad rolled off the tongue more easily. It was simpler and catchier. So four months later, my two vacant buildings became Greenwich Village's Quad Cinema.

Because we were the East Coast's first multiplex, publicity was no problem. The *New York Times, Village Voice,*

and *Variety* all called us. This was news. We also placed ads like the one shown above.

Just as with the D-Fuzz-It, and because the movie exhibition business was in such a slump, plenty of people told me I was nuts to build the Quad. But we were profitable from the moment we opened in October 1972.

Because we were so successful, we soon had imitators. Businessmen like my dinner party companion saw our attendance figures in the trades and started cutting up their theaters.

What's happened to movie theaters since then hardly needs to be explained: many large old theaters were chopped up into several screens, and newly built multiplexes sprang up in suburban malls and cities across America, growing to ten, fifteen, twenty screens, and more. Today, despite the fact that almost fifty screens now show movies within a few miles of the Quad, we still turn a profit, showing the best independent and foreign films. The Quad is a New York City institution and, according to former Mayor Ed Koch, "one of New York's best off-beat film houses."

> The man with a new idea is a crank—until the idea succeeds.
>
> —*Mark Twain*

Unfortunately, many great old theaters have been demolished. In addition, the new megaplexes often crammed as many people as possible into theaters the size of screening rooms for maximum profit. Neither of these developments makes me happy. Then again, many beautiful old theaters were able to stay in business by revamping themselves. And even the multiplex owners are figuring out that they can cater to the full range of moviegoers by showing blockbusters in large main halls and screening edgier or foreign films in smaller theaters. The movie exhibition business survived by innovating, and it grew healthier and more exciting because of the greater number of screens. The end result for movie fans, like me, is more choice. I give that two enthusiastic thumbs up.

As a postscript to the Quad story: Not long ago, a friend pointed out to me that two other men claim to have invented the multiplex. Recently deceased AMC Entertainment CEO Stan Durwood said he built the first two-plex in 1963 in Kansas City because structural issues in a building prevented the construction of one big theater. He followed up with a four-plex in 1966. Apparently, one James Edwards disputed this. He claimed to have invented the first multiplex (a two-plex) in Alhambra, California, in 1939. I have no interest in entering this who-did-it-first fray. What I know is that I hadn't heard about these theaters when I built the Quad, which was certainly the first multiplex in New York City, and by all accounts, the first multiplex on the East Coast. I'll take all credit—or blame—for that much.

What's Wrong with This Picture? Observation and Curiosity

As we've seen with the Quad Cinema and the D-Fuzz-It, the inspiration for inventions can come from just about anywhere. When people ask me where I got the ideas for a four-plex and a sweater comb, I tell them: At a dinner party and a dude ranch. But the keys are observation and curiosity.

Inventions solve problems. You can't see problems if you aren't observant, and you won't invent solutions if you aren't curious. People sometimes think inventors sit in a chair and get marvelous ideas out of thin air, but that has not been my experience. If I hadn't observed what that concrete wall did to my sweater, I'd never have thought, "I think I'll invent a device to care for sweaters today." And if I hadn't been curious after a chance conversation about the movie exhibition business, I would never have come up with the four-plex concept. When your ideas are prompted by observations and curiosity about the world around you, you stand a better chance of inventing things that other people will care about and need.

> The important thing is to not stop questioning.
>
> —*Albert Einstein*

I once had an employee who marveled at my insatiable curiosity. One day he commented that he figured I could probably improve the light switch on the wall, if I considered it long enough. I think he's right—not because I'm such a brilliant fellow, but because I love to think about such things. I can't understand boredom because there's always something to look at and think about. The how-does-that-work questions that used to drive my parents batty are the basis of inventive thinking.

I invented a simple game—which I am now using as a SKYY Vodka promotional item—after observing the way an empty yogurt container bounced when I missed the trash can and it hit the floor. I observed the bounce, tested it a few more times because it surprised me, and then studied the shape of the container. I imitated that shape in my playing pieces, and the result is a simple bounce game for one to four players.

When the theater owner began complaining to me about his family business, I could have nodded politely and then forgotten all about it. We were, after all, just making small talk and what did the movie theater business have to do with me anyway? But when I heard that theaters were empty, I was curious enough to wonder why. My informant didn't have a good answer, so I set myself the task of finding out. I knew that problems are opportunities for inventors—if we can figure out how to solve them.

When you encounter a problem, begin by asking questions, the more basic and naive the better: Where does it happen? Who is affected by it and cares about it? How does it happen? When did it begin? Why is it important? (Gelb 1998, pp. 67–68)

I did some research. Movie attendance had peaked in 1946 when two-thirds of the American population went to the movies at least once a week. In the 1950s, with the arrival of black-and-white television, the movie business turned to color, 3-D technology and Cinemascope to compete, but movie audiences steadily dwindled. By the 1960s, the movie-going population was only a quarter of its 1946 size. During the Golden Age of TV, a movie house manager put a sign on his theater door that read, "Closed Tuesday—I want to see Berle, too!" referring to Milton Berle's popular *Texaco Star Theater.* Clearly my theater-owning friend was right to rail against television. But I knew that TV wasn't going anywhere. As Federico Fellini is reported to have said, "To attack television would be as absurd as launching a campaign against the force of gravity" (Winship 1988, p. ix).

Just as talkies had lured audiences back into theaters during a mid-1920s slump in attendance, I reasoned that innovation could entice them back in the 1970s. I felt that the novelty and expanded viewing choices afforded by multiplexes were a way to hold onto and even expand the audience. As my Joe and Mary scenario showed me, and as my own movie-going experience confirmed, home entertainment could never replace the communal pleasures of seeing movies at a cinema, and there would always be a dependable ticket-buying population with reasons for going out rather than staying in.

Another aspect of the theater business that could have discouraged me is the fact that distributors are reluctant to give smaller theaters their biggest pictures. Instead of treating this as a handicap, I made it our strength. With several smaller theaters, you can afford to show films that will attract a more limited audience. Almost from the beginning, the Quad made art, film festival and independent films its specialty. (We'll talk more about niche marketing later.)

The conclusions I made while coming up w[illegible] idea for the Quad were based on observation and common sense. In fact, the four-plex idea seemed so simple and obvious to me that I couldn't believe industry insiders hadn't thought of it themselves. But people who work in a particular field become used to doing business in set ways. They may say they believe in innovation, but in actuality they often cling stubbornly to "that's the way we've always done it" thinking. As an outsider, you have the advantage of viewing situations with fresh eyes and fewer preconceptions.

In *Rules for Revolutionaries,* former Apple Computer, Inc., chief evangelist Guy Kawasaki calls this outsider advantage "harnessing naivete." He cites a great, possibly apocryphal, example from General Electric. In the 1930s, new engineers in the incandescent lighting group were welcomed with a practical joke. The initiation rite consisted of being assigned the "impossible" task of inventing "a coating for light bulbs that would remove the hotspot in the then current state-of-the-art design." No engineer was able to create this uniform glow bulb until around 1952 when a newbie did. He didn't know it was "impossible," so he wasn't trapped by a set of expectations (Kawasaki 1999, pp. 18–19).

> Common sense is not so common and is the highest praise we give to a chain of logical conclusions.
>
> —*Eli Goldratt,* The Goal

Author Denise Shekerjian warns against "the twin opiates of habit and cliché.... The more adept you are at something, the less likely you are to appreciate a varying interpretation...[or] generate new approaches" (Shekerjian 1990, p. 99). This was certainly true for movie theater owners in 1972. I didn't assume, like the veteran GE engineers, that since no one had come up with a solution, there must not be one. And while theater owners thought the only solution to their problem lay in drawing more people, I saw the solution might be more movies. Don't assume you have to be an expert or an insider to invent and innovate.

If you are observant and curious, and if you get a kick out of thinking, you've probably imagined ways to improve existing products that you use every day. Bank employee George Eastman didn't invent the camera, but in 1877, after buying the bulky state-of-the-art model and all its accoutrements, he did recognize the need to make picture taking less complicated. He pioneered paper-backed film, doing away with heavy, breakable glass plates, and he kept innovating until by the turn of the century, people around the world were taking pictures with his small, light Kodak box cameras (Newhouse 1988, p. 160).

> In the beginner's mind there are many possibilities, but in the expert's mind there are few.
>
> —*Shunryu Suzuki,* Zen Mind, Beginner's Mind

Many of my inventions have been improvements on existing technology: a more comfortable dental X-ray device, a safer cataract removal instrument, a more effective hypodermic safety needle. You know what they say about building a better mousetrap. Ask yourself the following questions about your improvement idea. If you can answer yes to even one of them, you may be on to something.

- Is it more environmentally friendly or durable?
- Is it less costly or time consuming?
- Is it safer or easier to use?
- Is it smaller or quieter?
- Is it more comfortable or attractive?

And consider these ten techniques for fostering your creativity and enhancing your inventing life:

1. Observe the world around you and be curious about what you see.
2. Study problems—think about why they exist, who they affect, and how they might be solved.
3. Ask questions, of yourself and others.
4. Look up words you don't understand and subjects that are unfamiliar to you.
5. Recognize and embrace the advantages of being a nonexpert.
6. Read as much as you can. (See the Inventor's Reading List on p. 29.)
7. Pay attention to surprises (fuzz balls on a concrete wall) and accidents (bouncing yogurt containers), and make connections—think of ways to use that grabbing or bouncing (D-Fuzz-It, SKYY Bounce Game).
8. Carry a notebook to record your questions, insights and reflections.

9. Talk with others. Seek out intelligent give and take in a small group, with egos checked at the door, and with mutual criticism that requires each person to defend his or her positions.
10. Go to places like San Francisco's Exploratorium museum of science, art and human perception—places where the imagination and ingenuity of others is on display and may spark your own.

➤ *Problems: Yours, Mine and Ours*

Observation and curiosity can lead you to unexpected and unforeseen places, including previously unfamiliar industries. But sometimes you need look no further than your own life. I wore sweaters all the time and I knew there was a need for a simple, inexpensive way to care for them; I wondered why alcoholic beverages gave me headaches and so I created a less-irritating vodka; I was so uncomfortable with the contraption my dentist used to take X rays of my teeth that I devised a more comfortable X-ray holder and target.

> Put an ingenious person in intimate contact with a problem and he or she will invent a solution.
>
> —*Anne L. Macdonald,* Feminine Ingenuity

The Baby Jogger idea came to a runner who wanted to be able to take his infant son along on his runs. He knew that if the device he created in his garage solved *his* problem, it might also be of use to others. Teva sport sandals were developed by a Colorado River guide to solve the problem of wet athletic shoes for river rafters. The sandals went on to become practical and popular for all kinds of outdoor activities (Thomas 1995, pp. 12, 174–75). Ann Moore and her mother, Lucy Aukerman, invented the Snugli soft infant carrier to satisfy Ann's desire to maintain nurturing closeness with her infant daughter while moving about (Macdonald 1992, p. 337).

One of my favorite problem-solving invention stories is that of the OXO Good Grips product line. Sam Farber, the founder of Copco, a successful cookware company, and his architect wife Betsey, were doing a lot of cooking together for their friends the summer Sam retired. They came to realize that none of the kitchen tools they were using met their needs. Betsey's mild arthritis made her especially aware that a hand tool is only useful if the tool-hand connection works well; Sam reasoned that there must be many people like them who, for one reason or another, and at one time or another, found handling kitchen utensils troublesome.

He telephoned Davin Stowell, founder of the industrial design firm Smart Design. Smart Design had previously done projects for Sam at Copco. Sam told him he wanted to develop a line of kitchen tools that would be comfortable, affordable, beautiful and dishwasher safe.

They tossed out the usual preconceived notions and started from scratch. Through consultations with experts and a long trial-and-error process of model making and testing (which we'll examine in greater detail in chapter 2), they crafted a small line of kitchen tools based on the principle of Universal Design—products designed to be comfortable and easy to use for people of all ages and abilities. Farber came out of retirement to start up OXO and make Good Grips™.

The distinctive swivel peeler quickly became a favorite best seller and OXO's signature product. Today there are more than three hundred products in the Good Grips line, and many were inspired by observed problems. Davin Stowell's messy experience of teaching his young daughter to bake cookies led to mixing bowls with nonslip bottoms. One Smart Design designer was a licensed weekend shell fisherman, so he created an oyster knife with a new blade shape.

Good Grips problem solvers have been honored with numerous awards, including the American Culinary Award of Excellence, the Good Housekeeping Good Buy Award, a Tylenol®/Arthritis Foundation Design Award, and has even been included in the permanent collection of the Museum of Modern Art in New York.

If your eyes and ears are open, as Sam Farber's were that day in his kitchen, you will notice how you or someone near you is compensating, making do, or fashioning a jerry-rigged solution to get something they need or want. Over the years, my interest in medicine and my conversations with a number of physician friends about the shortcomings of various medical devices have led me to create and license a number of medical instruments.

Remember: Problems are opportunities. Observe your own life and how others make their way through theirs. Let yourself be curious about the problems you see, and try to imagine how things might be better. Then arm yourself with the practical tools to make that potential solution a reality.

The Real World: Move from Idea to Invention

If I had to choose another career, I would teach introductory physics to nonscience students. I think I'd make a good teacher, and I'm sure I could get kids excited about the little miracles we take for granted—talking by telephone to people around the world, driving a car, switching on a light. Once kids understand the scientific principles behind these everyday acts, they see the world differently. Plus, when you understand how things work, you can start to think about improving them. I wish I'd had more adults around me when I was a kid who could answer my how-does-it-work questions, and I'd love to be able to excite kids about that kind of learning.

When I was growing up, one of the few adults to take my constant how and why questions seriously was the pharmacist I worked for as a teenager. I wasn't burly enough for the typical after-school job that most neighborhood kids did—boxing and delivering groceries at the A&P—so when I was twelve or thirteen, I started working for Saul, delivering prescriptions and helping out in the pharmacy.

One day I asked Saul how gold plating was done. He gave me a vague, preoccupied, adult kind of answer, but when I pressed him, he admitted he really didn't know. "Go to the library," he said. "Check out a book on electroplating and bring it to me. We'll read the book together and figure it out."

> The desire to know is natural to good men.
>
> —*Leonardo da Vinci*

These days when I have that kind of question, I'm likely to call Al Kolvites, the head of my product development lab. Al is an engineer, a commercial pilot and an experienced product designer and developer. I graduated from the Philadelphia College of Science and Technology. But nobody knows everything! If I ask myself a question that I can't answer, such as, "How the hell do they brass-plate objects? Brass is an alloy, how is it done?" I'll call Al and ask him. He'll say, "That's a damned good question," and set about answering it. Neither one of us may *need* to know the answer, but someday, in the process of fleshing out an idea, knowing how to brass-plate something might come in handy. If noticing and being interested in a problem is the first step on the way to an invention, being capable of solving it is the second. So we make it our business to cultivate a broad base of knowledge.

With few exceptions, inventions are *things,* not ideas—despite the fact that the U.S. Patent Office is issuing more and more patents for thoughts and ideas in cyberspace, such as Amazon.com's One Click feature. When you are trying to turn your idea into an invention, an understanding of materials and mechanisms is essential. Invention favors the prepared mind.

In the early stages of your product development, after you've pinpointed a problem and while you are brainstorming solutions, you might use a stand-in substance. Al calls it "nonexisteum." Nonexisteum is infinitely light, infinitely strong and costs nothing. With it, you are free to be very creative and your plans proceed beautifully. But at some point, you have to get real and find an existing substance that will meet your design's demands. What kind of adhesive will withstand high temperatures? Whic resist rust? What kind of fastener will keep manufactur costs down? What will make this game safe for children? That's when it helps to know how things are made and how they work—and maybe even how to brass-plate something. We wouldn't know who the Wright brothers were if all they'd done was say, "We have an idea for a machine that flies through the air. You can ride in it and look down from it and it'll be great." We know who they were because they tested their ideas and slowly learned the principles of lift and thrust, and how to achieve lateral control. They made their idea *real*.

Guy Kawasaki cites ability and passion as the most important traits of revolutionary innovators, like those who developed the Macintosh. He calls them "evangineers." They're evangelists who want to change the world, *and* they're engineers who have the technical knowledge to do it (Kawasaki 1999, p. 30).

> A man should keep his little brain attic stocked with all the furniture that he is likely to use.
>
> *—Sir Arthur Conan Doyle*

Acquiring this knowledge isn't that difficult. When making my first crude D-Fuzz-It prototype, my science background told me that aluminum oxide might replicate that concrete wall's grabbing properties. But when I started to think about the Quad, I had to educate myself about the movie exhibition business. I observed the crowds in theaters, studied the industry's trade journals and talked to people about what was important to them in a theater.

It's been my experience that people like to talk about their areas of expertise. If you've done your homework, ask intelligent questions and are honest about your motivations, most people, especially experts, welcome your desire to understand their field. There's no need to disclose your potential invention. You aren't asking these people to invent for you, you are simply asking technical questions about materials or processes you might use. If the Wright brothers had asked a cabinetmaker, "What's the best way to laminate and glue wood so that it is light and strong?" is the cabinetmaker inventing the airplane? Of course not, so don't be shy—question authorities.

To keep our knowledge current, Al and I read widely in a variety of technical fields. Read what interests you. Read to answer a question that pops into your mind. Read when a problem arises and you don't know how to solve it. Read until your vision blurs because when you invent, you are almost always building on and using previously discovered information. Al and I subscribe to trade journals and industry publications, many of which can be accessed for free at libraries or on the Internet. We keep up with what's going on with things like lasers, plastics and adhesives because we know that perfecting our next invention may require this kind of information. You should do the same.

Years ago, Al and a friend created a dimensional board game that they pitched to a major toy company. The game was called "The Spider and the Flies." A "web" sloped down to the spider's lair. Hidden magnets along the paths the flies took in their attempt to steal the spider's treasure randomly caused some flies to flip over and "die" before reaching their goal. The vice president of new products loved the game but ultimately passed on it. His research told him that the game would be too expensive to produce because of the intricate shapes and paths on the board. Today, with a great deal more knowledge about materials and the manufacturing process, Al could probably tell the

Inventor's Reading List

1. *Your local and a national daily newspaper such as the* Wall Street Journal *or the* New York Times
2. *Magazines such as* Forbes, Fortune, *and* U.S. News and World Report
3. *A wide variety of product catalogs (and you thought that was junk mail!)*
4. *Trade journals and trade association publications*
5. *Government studies, available from the U.S. Government Printing Office (Washington, DC 20402), such as* Survey of Current Business, Business America: The Magazine of International Trade, Official Gazette of United States Patent & Trademark Office of Commissioner of Patents and Trademark *and the Economic Report of the President (see the appendix for addresses)*
6. *The Internet, for company home pages and general research*

toy company how to produce it economically. Knowledge like that can turn a red light green. (Sure, today "The Spider and the Flies" would be unlikely to thrill a child accustomed to far more sophisticated computer games. But sometimes a low-tech game can still engage children *and* adults—see chapter 4.)

When you have a broad base of knowledge you are better able to make connections. The ability to make connections is a key component of creativity that can also be vital to inventing solutions. When Al worked at InterMetro Industries Corporation, he was assigned to a shelving project. His task was to offer the customer shelving made out of something other than stainless steel, which is a great material because it doesn't rust, but which is also very expensive. His team came up with an innovative composite plastic shelf, based on existing technology pioneered, in part, by the aircraft industry. They reasoned that if this composite construction was strong enough for an aircraft wing, it was strong enough for a shelf. Al was able to make this connection because of his interest in aviation and aviation technology. The resulting shelving won awards for innovation and made InterMetro a great deal of money. *Make connections.*

The OXO Good Grips Salad Spinner is the result of a trip to a toy store. The designers replaced the awkward cranks, winding mechanisms, and pull strings on existing spinners with a simple pump inspired by a child's spinning top. Again, a connection was made.

How could being a film buff lead me to get a patent called "The Use of LEDs, Flashing at High Speed, to Reduce the Consumption of Electricity in Stoplights"?

Because I'm interested in film and have had a lifelong fascination with vision, I knew that our eyes retain images on the retina for one-sixteenth of a second *after* seeing them. Motion pictures evolved from this understanding, set forth by Peter Mark Roget (the thesaurus fellow) in his 1824 paper, "The Persistence of Vision with Regard to Moving Objects" (Newhouse 1988, p. 171).

This fact led me to imagine the advantage of using a cluster of LEDs (light emitting diodes) in a stoplight. By flashing very rapidly they give the illusion of being on constantly, but they use substantially less power than a bulb that is on nonstop. I'm currently investigating the best way to market this invention.

Another example: Some years ago I was discussing cataracts with an ophthalmologist friend when he told me about a cryogenic cataract-removal procedure he regularly performed. Liquid nitrogen was used to freeze an implement that was then used to clear away lens fragments in the eye. I thought this sounded very dangerous—if there's a spill, liquid nitrogen can cause severe damage. I thought there had to be a better way and told my friend I'd think about it.

Elementary physics—what I turned to when my parents couldn't tell me how we got ice cubes in July—teaches us how a refrigerator works. Very basically, a liquid changing to a gas (evaporating) absorbs heat, making things colder. If you fill a chamber with a compressed gas then open that chamber and release it into a larger one, it will absorb heat.

My solution at the time was simple, though I wouldn't use the same materials today. Back then we weren't aware of or concerned about the ozone-damaging properties of compressed gas refrigerants, and so I filled a tube with Freon. A button push punctured the tube and the evaporating compressed gas made the tip of the instrument very cold. You could then touch the lens of the eye with this tip, freezing it and removing it intact.

My friend tried the model I made and thought it was fantastic. I sold the device to Alcon Laboratories, a company that specializes in ophthalmologic pharmaceuticals. They called it Cryophake and did very well with it. I could invent this device because I made the connections between elementary refrigeration physics and the need for extreme cold in a medical procedure.

Phil Baechler built a working prototype of his Baby Jogger out of a used stroller and an assortment of bicycle and other parts. When Baechler and his six-month-old son unveiled the creation at a 10K race, some racers looked at them with bewilderment, and others with envy. Baechler was asked where he'd bought the device, and he realized he could have begun selling them right there and then—if he'd had more than one. He first needed to figure out how to create a manufacturable product.

Baechler experimented with framing tubes from pieces obtained at an aircraft plant and with both steel and aluminum tubing. All presented problems. A visit to a boating supply store with a friend proved serendipitous. There Baechler found a plastic hardware connector used for boat railings that formed a strong yet flexible joint perfect for connecting aluminum tubing. Inspired by this, Baechler began designing his own connecting pieces using custom plastic molds, which eventually led to the first Baby Jogger patent and a product that was innovative, well built, and more easily reproduced. Learning about materials and making connections are crucial (Thomas 1995, pp. 175–76).

➤ *Reality Check: Reinventing the Wheel and Other Common Mistakes*

Another aspect of getting real, one that comes up throughout the invention development process, is being realistic about whether you've found a problem that is worth solving, and/or whether your method of solving the problem is actually better than a solution that already exists. As I've said, if your ideas are based on observation and curiosity about the world around you, the chances are good that the problems you tackle will be of interest to other people. But even when you set about solving a problem you've noted, it's possible to lose sight of the bottom line. To emphasize this common inventor pitfall, I offer this cautionary and somewhat embarrassing tale. Let's call it the Cottage Cheese Incident.

I recently complained to a friend about how lousy cottage cheese had become. The texture of the product was unappealing and I couldn't pronounce the names of half the ingredients on the label. My friend suggested I try organic cottage cheese.

I got on my scooter and drove to Whole Foods, my local organic supermarket. I picked up a container of organic cottage cheese, went to the checkout line, and when the container was scanned, I found the cost was something like six dollars. "What?" I said to the checker.

"That's the price, sir. Do you want it?" he asked impatiently. I paid, took it home, and, when I ate it, I had to admit it was delicious, just like what I remembered eating as a kid. But $6 for a sixteen ounce container?

I did some research and realized I could make this stuff myself. I bought a couple of gallons of milk and cultured it. I heated the resulting yogurt until it coagulated into little balls of cheese, and then I strained it through cheesecloth. And lo and behold, I'd made a terrific cottage cheese. What's more, I calculated that it only cost me about $2 for a sixteen ounce container.

I was quite pleased with myself until I realized that, given the amount of time I'd spent, I was working for about $3.50 an hour. This was clearly not a problem worth solving. Someone else had already solved the problem of making good, old-fashioned, additive-free cottage cheese. Factoring in my time, I really couldn't make it any more cheaply, so duplicating their efforts wasn't worth my while.

Then there's the story of Iridium, the world's first handheld global satellite telephone and paging service. When I first heard about them, the company was facing bankruptcy and looking for a buyer. They had launched a network of sixty-six, low-earth-orbiting satellites to facilitate uninterrupted communication around the world. But as *U.S. News Online* reported (8/30/99), their phones were "too bulky and too expensive" and often didn't work well. Setting up had cost them billions, but they only attracted a few thousand users.

I wasted a few dollars and a few hours on my perfect cottage cheese; Iridium appears to have wasted billions.

The more you have to lose, the more careful—in fact, the more ruthless—you need to be in reviewing your plans. Ruthless reviewing should precede and follow each of your actions. My Eleventh Commandment, one that reflects the unfortunate Achilles' heel of many inventors, is Thou Shalt Not Bullshit Thyself. Don't proceed past the idea and preliminary research phase until you are certain you have a viable invention. Always ask yourself the following:

- Is manufacturing my idea feasible? Do I have some nuts-and-bolts ideas about how to carry it out?
- Is my idea worth implementing? Will it offer unique benefits over other products?
- Is it a clear winner? Is the market big enough to create decent profit margins in the not-too-distant future?

We'll talk more about this kind of project evaluation a little later, and I certainly don't mean to throw cold water on your enthusiasm. I agree with essayist Sydney Smith who wrote, "A great deal of talent is lost in this world for the want of a little courage." Goodness knows that I grab hold of my ideas like a pit bull and can't be dissuaded from pursuing the ones I believe in. If I had given up every time someone said "Are you *crazy?*", I never would have developed some of my most successful inventions. But while you have to be passionate to bring an invention to fruition, you have to be dispassionate in assessing your ideas—before you spend a lot of time and money on prototypes and patents.

Think about marketing from the outset. Can you *sell* this beloved brainchild of yours? Remember, having a patentable invention is not the same thing as having a marketable one.

If you are confident that you've solved a problem worth solving, and that you've solved it economically and with clear value to potential consumers, pat yourself on the back. Now we'll get cracking on your prototype.

c h a p t e r   t w o

PROVE YOUR INVENTION/BUILD A PROTOTYPE

Ouch! The Kanbar Target and the ROLLOcane

About five years ago, I went to a routine dentist appointment and was told I needed X rays. The technician put a contraption in my mouth and told me to bite down and hold still. I tried to cooperate but wound up yelling, "Get that thing away from me!" It was heavy, cut into my gums, and was so painful that I left without getting the pictures taken.

A few days later I had a meeting with Al and his wife and partner, Florence. They were based in Pennsylvania at the time, and every few months I'd fly out from San Francisco and we'd meet in New York to toss around ideas and brainstorm. All I could talk about that day was my nightmare at the dentist's office. I described the device my dentist had used. It was stainless steel and cumbersome, and I told Al we ought to be able to design something lighter and more comfortable. We talked about how we might do this and sketched out some preliminary ideas.

Al is a very talented tool and model maker. He soon created a model that improved on what my dentist had used. Al made some machined components and wax molds for the film holder. He painstakingly worked with the plastic, hand-tooling our designs. The beam, which holds an aiming ring at one end, outside the patient's mouth, and X rays at the other end, inside the mouth, was relatively easy to construct, but the X-ray film holder required more trial and error. We tried various degrees of stiffness; some were too soft, some too hard. We went through four or five generations of our design, giving our samples to a dentist to try with his patients. The dentist would tell us if a part was too hard to move, if the bites were soft enough for patient comfort, what parts a dentist needed to be able to sterilize, and if the material we'd given him withstood autoclaving, or high-temperature sterilization.

We minimized the part count in comparison with the existing X-ray devices and wound up with a reusable aiming ring and calibrated indicator beam that can be autoclaved, and individually wrapped disposable bites. Because we used mostly lightweight medical-grade polymers in its construction, the tool was ultralight. We got patent protection and called it the Kanbar Target.

When we tried to license the Target, we were told that we needed to provide bites that could be autoclaved. But only hard materials—that defeated our purpose of comfort—can be used for that. So we decided to seek out distributors ourselves. When we were satisfied that we had a good design, we made a mold so that we could produce fifty samples and give them to dentists. We expected them to fall in love with the Target, place reorders, and spread the word about our wonderful innovation.

The Kanbar Target is an excellent product that meets all the criteria we set for it: it's light and soft and thus more comfortable for the patient; the ring and beam are reusable because they can be autoclaved and are thus economical for the dentist; it's adjustable and accurate, so it produces consistently good exposures. So why isn't it flying off the shelves? What happened to those reorders?

The Target solves the problem of patients who find other X-ray devices uncomfortable and sometimes even impossible to tolerate. But patients don't buy dental X-ray devices, dentists do. And dentists don't have a problem with the heavier devices. They may not even be aware of the discomfort of patients more stoic than me. What's more, our bites are meant to be disposable and autoclaving makes them soft and unusable. We found that dentists don't want to purchase disposable, single-use bites. The dentist who helped us design the Target thought it was great, but in retrospect we should have talked to more dentists. There simply isn't enough added value for the consumer—the dentist, not the patient—at least not enough to make them switch from what they currently use. Some dentists have even told us that they resent the whole concept of a target. They don't need help with their aim, thank you very much.

While I've always believed that it would be foolish to try to manufacture and distribute the medical devices I've invented and patented because I could never duplicate the reach and reputation of a Johnson & Johnson, I did think that I could handle distribution and sales in the dental arena. But we have had a hard time finding distributors and making sales. We haven't given up on the Target, but at the present time it is operating at a loss, something I'm able to accommodate because of my other successes. But all too often, inventors put all their resources into projects such as this, and if they lose, they lose everything.

Another cautionary tale is that of the ROLLOcane. As I watched my mother age, I began to see the need for older people to have a little bit of added support as they walked and shopped. It seemed to me that there were people who didn't require crutches or a walker, but who might benefit from a new and improved kind of cane. My goal was to offer rolling support instead of the jerky support of traditional canes and walkers, which must be repeatedly lifted and set down. A standard cane offers intermittent support, but a rolling cane would do so continually, increasing balance and promoting a more natural stride. I also knew that many people resisted using walkers because of their sterile, hospital-like appearance. I thought I could improve on that.

I got Al involved and we created an attractive, three-wheeled cane that provides support and security as you walk. It folds up for easy storage, you can steer the wheels smoothly, and it has hooks for hanging purses and a small platform for parcels. My mother likened it to a comforting "shoulder to lean on."

Once again, I made sure I met all the criteria I'd set forth for the product. We considered incorporating a reassuring braking mechanism, but we knew the cane didn't *need* brakes. The ROLLOcane has soft wheels, like those used on baby carriages, so it can't roll away from a user. Satisfied, I rushed the product into production, convinced we had a winner.

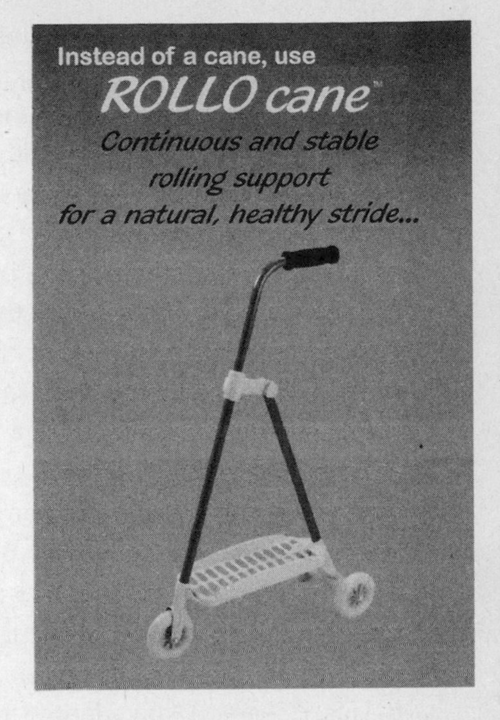

I was wrong. Like the Target, the ROLLOcane has enjoyed modest sales, but it has not been embraced as I thought it would be. Medical professionals object to the fact that its height is not adjustable, and some wish that we'd included brakes. Despite the fact that it is perfectly safe and won't roll out from under a user, the wheels frighten some people.

We should have made prototypes, taken them to nursing homes and physical therapists, and listened long and hard to the feedback we got. We didn't do this. Market research would have backed up my strong belief that there was a market for a rolling cane, but product research would have told me that to be better than what currently exists, the cane would need to be adjustable and perhaps have brakes.

We haven't taken the time to revamp and either relaunch or license the ROLLOcane and the Target because we have so many other front- burner projects. If these were my only inventions, I would modify them—and I've been told the cane is perfect infomercial material—but with so many other things going on, we don't have time to do much except sell through our current inventory.

Both these stories highlight important lessons inventors must learn if they want to find not just monetary success in the marketplace, but an eager, appreciative audience for innovations. First, correctly identify the customers for your product, and understand exactly what they want and need and how they'll use it. And second, build a prototype and test your invention with its potential users so you can identify mistakes and pitfalls and correct them before you take your product to market.

Two Popsicle Sticks and a Coat Hanger: Prove Your Invention Cheaply and Simply

I have an inventor friend who says that if you can, you should prove your invention with nothing more elaborate than two Popsicle sticks and a coat hanger. The idea for folding aircraft wings, so important for saving space on an aircraft carrier, was conceptualized using a bent paper clip and a rubber eraser. In other words, your first model should be inexpensive and rudimentary. In some cases, in order to prove your idea will work, or to get financial backing or a licensing deal, draftsman-quality drawings will suffice. And remember, the Patent Office doesn't require a working model.

Depending on your invention, you will probably develop a succession of more elaborate and accurate prototypes as you refine your ideas and test them with end users. Prototyping typically goes through some or all of these phases:

Basic Design. In working out the general shape of your invention you may wind up with something nonfunctional or something that functions crudely. This early design probably won't be made of the material you ultimately envision, as was the case with my homemade D-Fuzz-It.

Functional Model. This one is tested until it works as you'd like (the various Kanbar Targets generated in our lab).

Model Modified Based on Testing. This is tested on end users and improved.

Working Prototype. Ideally, this is tested by actual consumers in a real-world (not laboratory) setting (for example, the Targets we gave to a dentist).

Production Models. These are sold in a test market.

(Adapted from DeMatteis 1997, pp. 32–33.)

Many basic or merely functional prototypes can be made at home. By this I don't mean that you should learn to work with injection-molded plastics or metal casting in your garage (although to some extent you can); I mean that you should not invest thousands of dollars paying a manufacturer to do these things in your earliest prototyping phases. To begin, you need to prove to yourself that your idea is feasible, and depending on your end user and market, you may want to create something you can test out informally on a few sample consumers. Not every invention lends itself easily to a rough working model: I couldn't make a tiny Quad cinema and run a few films to see if people showed up— though I have read that in developing their Courtyard hotel chain, the Marriott Corporation built a prototype room with movable walls in order to test various possible designs (Thomas 1995, p. 214). Extremely technical or idea-based inventions are difficult to model effectively without creating a nearly final product.

When starting out, you want to keep things as simple and inexpensive as possible because you don't want to throw good money at model makers (or manufacturers or building contractors) for the execution of a bad idea. Injection-molding plastic, for instance, can be quite costly. If you spend a small fortune perfecting a finely polished first model, you are more apt to forget the Eleventh Commandment—Thou Shalt Not Bullshit Thyself—and ignore the complaints and red flags that come up in its testing. You may become blind to its faults. Just as hundreds of patents exist for unmarketable inventions, countless models of valueless ideas exist. If you spend thousands on a production-tooled model and then find it needs substantial modifications, you may have to scrap that expensive model and begin again.

If Popsicle sticks and coat hangers won't prove your invention, consider using clay, wood, plastic, fabric or vinyl, sheet metal or even cast metal.

Clay is great for making molds for a plastic or metal model. There are also modeling materials in art supply stores that are pliable, like clay, yet hard and durable after being baked.

Wood can simulate plastic or metal. Balsa wood (available in hobby shops) works well if you need to shape parts. You can ask for help, or find an inexpensive helper, at a woodworking or model airplane shop.

Plastic? You *can* try this at home. It's not brain surgery—bending plastic over an electric stove is easy—but you do need to educate yourself about techniques, materials, tools and safety. Plastic suppliers, hobby shops and art supply stores carry liquid plastics that can be molded in simple handmade molds, and sheet plastics that can be cut, bent and glued. Visit some stores and ask questions.

Many fabric or vinyl models can be made with a home sewing machine.

Sheet metal is relatively inexpensive and can be used to demonstrate something that will ultimately be made of plastic. If you take a crude cardboard mock-up to a sheet-metal fabricator, he or she can create a good, cheap prototype.

Cast-metal objects can be simulated with custom castable plastics. Kits and compounds are available from outfits such as Castolite (see appendix).

Many models require some machining—drilling, punching, shaping, planing and so on. If you have the equipment, you can do this yourself. If not, take your model to a machine shop or enlist the help of a hobbyist.

Some Prototyping Resources

- *Builders of model airplanes and other hobbyists can talk to you about the technology you need, or perhaps make your prototype at a low price.*
- *California Manufacturing Technology Center is a state agency that helps inventors develop their inventions.*
- Fine Scale Modeler, Design News *and* Machine Design *magazines for articles and supplier ads*
- *McCaster-Carr Supply Company (Los Angeles, CA), for a staggering assortment of tools and hardware*
- The Modelmaker's Handbook *(Alfred A. Knopf)*
- *Stock Model Parts sells an assortment of small parts and prototype kits.*
- *United States Plastic Corporation sells plastic materials, glues, and tools.*

For details on all of these, and other resources, see the appendix.

If you know of a part or product that is similar to what you want to create, consider contacting the manufacturer to see if any old or discontinued molds are available. You might be able to have that mold machined to suit your need. Cannibalized parts—handles, nozzles, switches and the like—taken from existing products can often work beautifully. These can be obtained cheaply at flea markets, garage sales and discount stores (Debelak 1997, pp. 133–45).

If you find you do need professional assistance, there are still ways to get low-cost prototypes. If you can get orders for your invention using drawings, a homemade model and good salesmanship, you may be able to convince a manufacturer to pay for a final working prototype in exchange for your future business. You will need to sell the manufacturer on the merits and moneymaking potential of your idea. One place to track down and talk to manufacturers is at trade shows.

At local businesses, you can seek out an industrial designer, machinist or model maker willing to moonlight and build your prototype at cost in return for a share of your future profits. And consider contacting the design department of your local college. You might find a student or professor you can employ cheaply.

The Good Grips people produced hundreds of testing models by carving Styrofoam and wood. In developing their salad spinner, simple 2-D sketches and 3-D volumetric foam mock-ups were used. Dimensionally accurate models were achieved with computer-aided modeling processes. (You can take a look and a spin at: www.oxo.com/comfort/spinner2.html.)

I invented a surgical instrument to be used by doctors in treating varicose veins—smaller, less invasive than the state-of-the-art technology at the time—and licensed it to Johnson & Johnson. After crafting a handmade sample, I invested in an inexpensive aluminum mold that could make a number of samples, so that I would have a working prototype to show potential licensees. (A more expensive steel mold could have made many more pieces.) For reasons we'll explore in greater depth when we discuss licensing, I've found a finished product to be very advantageous when pitching, much more persuasive than even the best drawings.

Because I had made a mold, when Johnson & Johnson asked for them, I could easily provide 150 samples. They test-marketed these samples, getting a great response and paving the way for our deal. The extent to which you develop your invention—whether you stop at precision drawings or continue on to production models—will depend on many things. What is customary in the field in which your invention will be competing? Can you license with drawings or a crude model? Can you write an accurate patent application before having developed a working prototype? Are you manufacturing and distributing on your own, or is licensing your goal? How many building skills and how much development money do you have? There are no hard and fast, across-the-board rules—except the need to prove that your idea can be turned into an invention that is marketable.

Finally, because it bears repeating: Keep it simple. This pertains both to making prototypes and to your ideas. The best inventions promise and exemplify simplicity in their function and design. What could be simpler than the Chip Clip? The iMac's easy set-up is a key selling point. My friend Knud Dyby was told by an injection molder that his invention was too simple to warrant a patent. He had taken a plastic box and put a circular magnet on the hole at the top and used it to store and dispense paper clips. The elegant simplicity of his idea and its execution could have made him millions if he had sought a patent. There's no such thing as too simple.

> Making the complicated simple, awesomely simple, that's creativity.
>
> *—Charles Mingus*

►Money, Money, Money, Money, MONEY!

While your first step after getting an idea shouldn't be mortgaging the house, some things do require cold hard cash. It's not true that you need a lot of money to make a lot of money —having resources can sometimes make it easy to be wasteful and imprudent—but, of course, having ready access to development funds doesn't hurt. Most of us, though, need help funding the prototype for our invention and, more importantly, funding its manufacture. Getting this start-up capital can be a daunting task and is almost an art in itself. It's beyond the scope of this book to describe fully the complicated world of raising money, but I can point you in some solid directions.

At this point in the development process, many people are attracted to those inventor's services companies that promise to take care of all kinds of bothersome details for you so that you can sit back and wait for royalty checks to arrive in the mail. Beware.

Invention assistance outfits rarely provide real help. It's in their best interest to tell you they have "researched" your idea and found it to be a guaranteed moneymaker. By doing so, they hope to persuade you to further invest in their "services." They will find you a model maker all right—if you send them a fat check. (Services that insist you first get a patent and then offer to present your invention to a variety of manufacturers are more likely to be legit.) You can't do everything yourself and you don't have to, but you should take charge. Learn to do what you can and learn all about what you hire others to do so that you can employ them intelligently.

When you do need to raise money, there are a number of ways to go. When a plastic mold maker said he wanted $1,200 to make a mold and prototype of my D-Fuzz-It, I went to friends, family and business associates and offered them a piece of the pie. If they provided start-up funds, they could get in on the ground floor and take a percentage of the business. Pity for them, most said no. But I did ultimately go into business with one friend who invested some initial packaging and marketing funds. Many, if not most, inventors find they need to make similar deals in order to realize their plans. The less developed your idea, the bigger the risk

> Money helps, though not so much as you think when you don't have it.
>
> —*Louise Erdrich,* The Bingo Palace

your partner or partners are taking and thus the greater percentage you're likely to need to offer them.

When I wanted to produce and market SKYY Vodka, I had adequate personal financial resources to do so. But I honestly believe that if I were twenty-nine years old, with the SKYY idea and no money, I could still have launched the product by giving 40 or even 50 percent of the business to investors. If you're a lousy pitch person, hire someone to pitch your product to investors or offer the pitch person a percentage of future earnings. When I was getting ready to launch my Tangoes game, I happened to meet a young salesman, Mark Chester, at a party. We had a great conversation and I was impressed by his hard "soft sell." I told him about my game, asked if he'd like to try to sell it in area stores, and he gave it a shot. Mark did well and said he'd like to represent the product. I said, "Fantastic!" because I wasn't interested in making sales or running the Tangoes business. I told Mark he could have the first $20,000 the game brought in and offered him an 80/20 split from there on out. We went on to become partners in the successful Tangoes business.

In raising money, as in inventing, you are only limited by your imagination. Approach the task as you would an inventing problem. Be creative. If you have a novel idea about how to approach someone or what kind of deal to make, don't dismiss it outright. Think it through, protect yourself with simple contracts and legal advice if need be, and then go for it.

There are almost infinite variations of the classic I-invent/you-put-up-the-money deal. As mentioned earlier, you can often trade a small percentage of your future business for designing, tooling or manufacturing services. Distribution channel financing is another option: you get up-front money for signing a distribution agreement with a larger company.

Baby Jogger inventor Phil Baechler and his wife, Mary, kept their day jobs and went through their $8,000 personal savings while building custom strollers in their garage. Though their strollers were selling, they had trouble finding investors—but they didn't quit. With a "patent pending" and the last of their cash, they placed an ad in *Runner's World,* incorporated Racing Strollers, Inc., and started selling mail order. Happy customers and free publicity generated by Phil's press releases helped the company grow quickly. By the time the patent was issued a few years later, sales were impressive enough for them to finally secure an operating line of credit and a development loan from the Small Business Administration (SBA) to build a real factory. You can make use of the SBA as well, and your local chamber of commerce should also be able to direct you to local business assistance offices.

What about banks? Banks will lend you money if you have a track record and they believe you can pay them back. Lending money is their business, but banks are inherently conservative. And backing yourself up on paper for a bank can be difficult. It's harder to write a convincing business plan for new inventions because you are dealing with something new and there are many unknowns. Obviously, you have reason to believe in the sales potential, but banks generally want much more hard data than you can provide. Banks don't want to be venture capitalists, and most small businesses can't meet their tight loan requirements.

Some Financing Resources

- *National Association of Small Business Investment Companies*
- *The Small Business Development Center (SBDC) nearest you (for state sources of funding) and the Small Business Administration. Your local Certified Development Corporation can help you prepare your application for government-backed loans.*
- *Small Business Innovation Research awards grants for areas of interest to the Departments of Defense, Health and Human Services, Agriculture, Transportation, Education, and the Interior; NASA; the EPA, the DOE; the National Science Foundation; and the Nuclear Regulatory Commission. Requirements are rigorous.*

See the appendix for contact information.

And what about those venture capitalists? Are they all just New Economy whiz kids? Not at all. Anyone can be considered a venture capitalist if they give you money for a share of your business. They can be friends or relations, business associates, local businesspersons, former employers or angels—that is, private, part-time investors.

The New Economy has indeed bred more and more venture capitalists, and there are all sorts of ways of finding them. Some resources include:

- *Pratt's Guide to Venture Capital Sources*
- Publications such as *American Venture Magazine* and *Entrepreneur*

- FinanceHub.com (www.financehub.com), information and resources for entrepreneurs and investors.

Management consultant Edwin E. Bobrow points out that "venturing is the hardest way for an inventor to launch a new product." Licensing is much easier (and is discussed in chapter 4). Bobrow also stresses that venture capitalists look for the promise of "great management." They "invest in people, not technologies and not products" (Bobrow 1997, pp. 242, 244). This is true, in one form or another, for anyone who might become your partner in business or an investor in your invention.

While an excellent product is a prerequisite, the way to convince people that you know what you're talking about and have the necessary organizational and management savvy to pull it off is to create a business plan, a well-thought-out, researched and documented case for your product and its market. In your business plan, you need to evaluate your product realistically, assess its market and market appeal, and project potential market share, costs and sales. It's like building a working prototype of your invention, and just as vital. You write a busi-

Business Plan Resources

- How to Write a Business Plan *(Nolo Press)*
- *The Small Business Administration*
- Your First Business Plan *(Small Business Source Books)*

See appendix for more information.

ness plan to prove to yourself, and then to others, that the product is real, that it's worth producing and that you'll be able to sell it to someone, preferably lots of someones. Just as the prototype tests your idea with potential customers, the business plan sells it to potential investors. And whatever else it contains, it must convey your viability as a smart manager, organizer, planner and thinker. Let's look at how to assemble this proof.

People Watching: Observe the End User

To prove that you have a launch-worthy product, you need to conduct clearheaded research of your market and ruthlessly review your invention at every step. Depending on your product and your personality, you might want to do the bulk of this analysis and research at the beginning, when all you have is an idea, or you might wait until after you've built a basic prototype, but it must be done before making large financial outlays. If you're funding your own efforts, don't think you can skip this process, as it's vital to your ultimate success. And if you're looking to others for start-up funds, anyone except perhaps an indulgent relative with deep pockets is going to expect you to gather this data and put it into a coherent, convincing document.

Al's previous employer, InterMetro, had an almost unbelievable success rate for their new products. Something like nine out of ten of them went on to make money. The national average is one out of ten. What made them so special?

They followed a process, very much like the Five Fundamental Steps in this book, and made certain that market, consumer and design needs were aligned. Marketing defined the need for a product, proving that a new product was needed to fill a void or that a much improved version was needed to meet and beat the competition. InterMetro's industrial design department researched and defined what the product had to be to meet or exceed the customer's expectations. The engineering department worked with industrial design to ensure that the product could be profitably manufactured while meeting consumer needs. Their constant vigilance to reconcile all of these needs ensured that the company was never in the position of trying to *convince* customers that InterMetro had produced what they wanted. Instead they produced exactly what customers needed.

To find out what the consumer wants and needs, do market research. And by this I don't mean focus groups and the kind of pie charts and graphs consultants charge a small fortune for. If focus groups worked, the big companies would never put out products that fail, which they do all the time. Guy Kawasaki faults focus groups because unlike real life, they "are clean." In them, "participants project how they would use a product—sitting comfortably in a room with other people listening to them, with conversation facilitated by a professional, and feeling like they have to express an opinion since they've been paid to be in the group" (Kawasaki 1999, p. 50). My theory is that focus groups are used by executives to cover their tracks—when their products flop they can slap their hands to their foreheads in bewilderment and say, "But we had the numbers!"

Kawasaki also says that "since nothing is more important than gathering information about your customers . . . you should never leave it to marketing research professionals." He contends that their methods are unable to detect and communicate subtle findings, they miss unforeseen opportunities, they provide stale information because of the time it takes to make all those pie charts and as a consequence, they let crucial issues fall through cracks (Kawasaki 1999, pp. 115–16). You must see for yourself, on the customer's turf, what the problem you think you are solving really is.

Much of the market research you might pay for will give you composite portraits of potential buyers based on averages. But averages that are supposed to please everyone usually wind up pleasing no one. I was hosting a dinner party one evening, and when it came time for coffee, naturally some guests asked for regular and some wanted decaf. After a sleepless night, I queried my housekeeper about her coffee preparation the night before. She explained she had brewed half a pot of regular coffee, half a pot of decaf, and then mixed them together. "I thought half and half would make everyone happy," she explained. I wasn't 50 percent happy, I was 100 percent annoyed!

So what kind of market research do you have to do? It really comes down to talking to and observing your end

Low-Cost Market Research

- *Test your reasoning, not necessarily your idea, on people you know will be absolutely honest with you. ("Would you buy a gadget that quickly and easily removed sweater pills?" not "How would you feel about a sweater comb?")*

- *Approach people who work in the market you are trying to enter and offer to pay them for a few hours of consulting time. Again, you need not disclose your idea. Ask about existing products, consumer feedback, and any unusual sales factors.*

- *Talk to distributors, store owners and salespeople.*

- *Go to trade shows.*

user, your potential customers, and assessing the market they create.

The phrase "the customer is always right" is not an unfounded truism: customers vote yea or nay with their dollars. If they vote nay for any reason—a single feature of your product is discouraging or unwanted, the product looks undistinguished on the shelves, the product isn't sold in the places customers expect to find it—only you and your invention lose. You must involve them in your process, learning more about them than even they know about themselves.

Begin by going where your end users live, work and shop. If you'd like to see your product sold at Wal-Mart, go to Wal-Mart. Talk to people buying items similar to yours. They are the ideal focus group. Talk to the people who sell items similar to yours; they can give you information on what sells well for them, and what might sell even better. And remember my experience with the Kanbar Target: make sure you correctly identify who your real customer is. In some cases, the ones who buy your product are not the same ones who use it directly, and you need to please the people who buy. If you hope to get picked up by a distributor, also talk to distributors; you need to convince them first that they want your product before you will have a hope of convincing the end user.

Some inventors get nervous at the mere thought of talking about their invention to anybody, much less with lots of people. "They'll steal my idea," they think. The fear of having your idea stolen must not outweigh the need to test your product out on people. In reality, the chances of someone stealing your idea are much lower than the chances of making an inferior, doomed product because you didn't talk to your customers first.

You've got to consider the opinions of store buyers as well—if *they* can't be convinced of the uniqueness of your product, you'll never get it on the shelf. I designed the sleek, easy-to-use SKYY Timer as a promotional giveaway item. People who used this small, digital, magnet-backed timer found it terrifically simple to use and handy. I explored the possibility of selling it as a stand-alone item, sans the SKYY logo, but retail buyers didn't see the need for another timer among the many in the marketplace, and weren't interested in stocking it. The timer solved my problem of needing a nice SKYY promotional item beautifully, but the retail world didn't perceive a customer need for it.

Not that giveaways can't be turned into great products. S.O.S. steel wool soap pads were created by door-to-door salesman Edwin W. Cox as a way to get his foot *in* the door to sell kitchenwares. They quickly became more popular than his pots and pans. Same with Avon, which began when door-to-door book salesman David McConnell started to give away small bottles of perfume to people who listened to his pitch. Unfortunately, I should have assessed the timer market and consumer much more thoroughly before deciding to package and sell my giveaway (Panati 1987, pp. 102, 243–44).

Also, it's not enough to talk to people; you have to watch them buying and using similar products. Often, end users can't articulate what they want, but if you watch them, you can recognize their needs and offer them something they didn't even know they needed, something that will prompt them to say, "How did you know? That's exactly what I could use."

For example, at InterMetro, the product design and development department was given the task of designing a new dish dolly, the piece of equipment that holds dishes in stacks in restaurant kitchens. The industrial designer stood in the background and watched the dishwashing process, which in large establishments is done by people and machines. He tried to blend into the background, since if people know they are being watched, they tend to work more conscientiously and mind their manners.

> Better to ask twice than lose your way once.
>
> —*Danish proverb*

The designer found that the dishes weren't treated very carefully. Workers would bus the dishes, scrape off food scraps, and then load the plates at the front end of the dishwashing machine. Sometimes they'd play a bit of a game with the person unloading dishes at the other end of the machine. The front-end folks would load more dishes on the machine than the unloader could unload. The dishes come off the machine very hot and very fast and need to be stacked. If the machine is overloaded, some poor guy ends up with a huge, unwieldy stack that he literally drops into the dish dolly like a hot potato—not unlike the candy-making episode of *I Love Lucy.* He can't take the time to gently deposit one dish at a time, and in any case, his fingers won't fit inside the round, plate-shaped hole at the top of the dolly. You can guess the result: lots of cracked and chipped dishes.

The design team introduced a dolly with slots on the sides so that kitchen workers could lower their hands in and deposit, rather than drop, stacks of plates. When they introduced the product at a trade show, competitors were green with envy.

This simple design feature was not the result of consumer requests. When asked what they needed, dishwashers said, "I don't know. Everything is fine as it is." The new design was the result of careful observation.

Assessing the wider, or even national, market is every bit as important as polling individual consumers. I've ridden scooters for years. I love their maneuverability and the way I can park where car drivers cannot—quite a plus in my New York City and San Francisco home bases. People are often surprised to find out I've driven a scooter to dinner or a meeting, but I usually beat them to our destination!

My manufacturing and design company, Rex Products, Inc., is considering an alliance with Bajaj Auto Ltd. of India. Bajaj has a fabulous, inexpensive line of attractive scooters, and we may be working together to bring them into the States. My love of scooters, my frequent travels in India and my friendship with Bajaj owner, Rahul Bajaj, have laid the groundwork for this alliance. But I wouldn't enter the scooter business for personal reasons. It would be because I've assessed the U.S. market and think our ideas are viable.

We've looked at the current and potential stateside scooter market from every possible angle. We've factored in traffic patterns, inadequate public transportation systems, parking limitations, and the number of younger consumers. We've explored the attributes and weaknesses of the competition and think we can compete with both the practical, low-cost Hondas and Yamahas and the stylish, high-end Italjet, Aprilia and Piaggio models. Because we believe that most consumers are seeking an environmentally sound, cost-efficient, user-friendly and stylish means of local transportation, we will be looking at targeting several niche markets: college students and young working adults. We know that additional markets, such as rentals and fleet sales, will emerge as 60cc–80cc, four-stroke, automatic transmission scooters become available.

We've noted that many college towns and campuses are similar to scooter-friendly European cities in that they are compact and have limited parking. We've learned that the number of college-age students will continue to grow over the next decade, so it would make sense to place dealerships in high-enrollment college towns. In addition to being a prime target market, college students are trendsetters for the rest of the nation. Based on our prime consumer and the attributes of our scooters, we will also focus our launch in certain influential urban communities. We've projected sales and mapped out advertising, marketing and promotion plans. We've done all of this *before* putting one scooter on the market.

When assessing your market, don't be afraid of targeting a niche, which is simply a small, discrete segment of a large market. I don't assume I can get every able-bodied American onto a scooter, but I do think I can sell our scooter to a significant segment of scooter buyers. Bob DeMatteis reminds us that giants like Apple Computer, Honda, and Hewlett-Packard each began by targeting niches in existing high-volume markets. "Virtually every entrepreneurial start-up effort begins as a niche" (DeMatteis 1997, p. 44).

I believed there was a segment of the alcohol-drinking population that would embrace a "cleaner" vodka. I was right, and now our market for SKYY Vodka knows no niche boundaries. If you aim your product and marketing efforts at a relatively small segment of the entire population, you actually increase your chances for larger success. Keeping your focus smaller pays off in several ways: it forces you to specify your appeal, it keeps the bar for success at a more attainable height and it limits your losses if you're slightly off.

Starbucks is another example of a company that successfully exploited its niche first before growing into a national force. In fact, when Starbucks began in 1992, it entered a small and even declining coffee-drinking market. In 1962, 75 percent of the population drank coffee, but thirty years later only 51.4 percent did so. (Sounds like the movie-going figures in the seventies, right?) So what was Starbucks thinking? Well, they had done their research and figured out that within that declining market, sales of gourmet coffee were actually growing. They had identified a niche, and an expanding one at that. Starbucks mastermind Howard Schultz didn't try to market to the masses; instead he offered "sophisticates" a superior product. The rest is history (Thomas 1995, pp. 20–28).

► *Play by the Rules*

The data you accumulate becomes your road map. The needs people express to you or that you observe become your product priorities, and the dimensions of the market determine the rules of your game. These rules and priorities become your Ten—or Twenty—Commandments. You break them at your peril. Gathered coherently in a business plan, they prove to potential investors and partners the viability of your invention and its market—and not incidentally your own determination and business savvy. Followed wisely, they more often than not will lead you to success in the marketplace.

Even if you think what people ask for is nutty, you'd better take their requests seriously. Al once worked on a low-pressure steam cooker for professional kitchens. Chefs told him they needed to be able to see inside the steamer while items were cooking. Al didn't get this at all—when cooking with steam, a cooked pea or kernel of corn looks exactly like a partially frozen one. But chefs are accustomed to being able to look in ovens and under lids while cooking. So Al struggled to find a way to put a window in the steamer, which drove him crazy because, of course, windows steam up.

Al eventually worked around the problem. He patented a power-conserving steamer that used only as much power as warranted by its contents. A light let cooks know when cooking power was being called for and when the machine was finished cooking. Reassured by the light as to what was going on inside at any given moment, the cooks no longer felt the need to see inside the steam cooking chamber.

As we've seen, Good Grips' ergonomic designs and innovations are based on extensive research and interviews. In designing their first product, the swivel peeler, they talked with consumers, chefs and retailers, and studied competitive products. After reviewing their research, they decided upon their necessary criteria: a handle large enough to grip firmly and avoid strain, oval-shaped to prevent the tool from turning in the hand, with a rounded end that fits comfortably in the palm and evenly distributes pressure, and an oversized hole for easy hanging. Meeting all of these criteria paved the way for their success and set the standard for the development of their more than three hundred ensuing products.

My Kanbar Target and ROLLOcane are both examples of the pitfalls of inadequate and incomplete research. For the Target, we did heavy product research but not enough market research; though we created a good product, it was a product aimed at the needs of the wrong person. With ROLLOcane, again, our product design was sound, but we failed to take adequate note of end users' perceptions. Logically, we knew brakes were unnecessary, but consumers would have found them reassuring. There really is no way around this process: you need to learn what your product and market commandments are, and then you have to follow through on every one.

Some products have more requirements than others. We couldn't simply import Bajaj scooters into the U.S., we would have to make them a viable product for the U.S. marketplace. This involves everything from adapting the scooters to meet stringent Environmental Protection Agency guidelines—pollution regulations we believe in and embrace whole-heartedly—to improving the look of the control panel and making the scooters available in popular colors.

In general I've found that most people don't argue with the need for this kind of research. The hard part for most inventors comes sometime after they've gathered the information and then realize that they must modify their invention. As difficult as it sometimes is, that's the goal of all this research: to get your rude awakenings *before* spending the money to launch. It can be humbling, infuriating or just plain discouraging to return to the drawing board and rethink your product over, and sometimes over, again. This development and prototyping phase can often take a long time, and this is good. It provides you with opportunities to discover unforeseen improvements. Creativity is an ongoing process of learning and adjusting; it's more than just a lightning bolt flash of inspiration.

As you build and test prototypes, you will more than likely have some disappointments, but you could also encounter happy surprises. For instance, Ivory soap didn't always float. Too much air was accidentally whipped into a batch. When consumers raved about the bar that couldn't be lost because it bobbed to the surface, extra whipping became part of Ivory's manufacturing process (Panati 1987, p. 218). Resist the temptation to gloss over inconvenient doubts or unexpected information in order to expedite your design process. Most product entrepreneurs rework their ideas repeatedly. And if your ruthless review uncovers the unpleasant truth that there is no viable market for your product, move on to your next project.

Recently, it made sense for me to capitalize on the success of SKYY Vodka by extending the line with a flavored vodka. Friends, customers and colleagues were in fact urging me to do just that. But I resisted doing so until I was satisfied we could offer something better and more unique than the other flavored vodkas on the market. That took time. We spent almost a year perfecting the SKYY Citrus formula. Getting a distinctive taste, one that couldn't be simply reproduced by a bartender with a lemon or lime, took time. We achieved our goal by experimenting with and ultimately melding lemon, lime, orange, grapefruit, and tangerine. But some early versions were cloudy. More adjusting, more experimenting. It took perseverance and determination to get everything—flavor, appearance and finish—just right, and it took a willingness to wait, trusting that launching an inferior or derivative product early was not as important as launching an excellent and unique product a little later.

> Next in importance to having a good aim is to recognize when to pull the trigger.
>
> —*Elmer G. Leterman*

My new Vermeer Dutch Chocolate Cream liqueur followed a similarly long developmental path. About three years ago, I tasted a popular cream liqueur made with Irish whiskey. I thought I could make a better cream liqueur with premium vodka and chocolate. Of course, I had plenty of experience with premium vodka, but finding the right chocolate proved difficult. When I found the best, a wonderful Dutch chocolate, it was quite costly and I was advised against it. But the Dutch have been making an art of fine chocolate for centuries and deserve their reputation for producing the best in the world. I stood firm and insisted on it. Then came a long—and not unenjoyable—process of creating and taste testing various blends. Our end result was worth every bit of time we took to perfect it. The time and trouble you take up front will save you both in the long run. Get it right the first time.

Even if you realize you've made a mistake after going to market, you can still sometimes recover—and learn invaluable lessons in the bargain. Look at General Motors' Saturn; they turned a mistake into a triumph. Despite years of research and development, Saturn initially faced production problems, a recessionary market and even significant product recalls. But what do you think of when you hear Saturn? Customer satisfaction, right? They turned their problems into opportunities. Promptly and honestly identifying and solving problems came to symbolize Saturn's commitment to having happy customers, it became part of their marketing, and it earned them high marks in the J. D. Powers and Associates 1992 survey of customer satisfaction (Thomas 1995, p. 15).

> I failed my way to success.
>
> *—Thomas Edison, inventor and holder of 1,093 patents*

If you have an invention that you know other people need and want, that you can make profitably and that has a defined market, it's time to protect it.

chapter three

PROTECT YOUR IDEA

Better Safe Than Sorry: The Needle Protector

A few years ago, my physician, Martin Sturman, showed me a design he'd come up with for a safety needle. Martin is a longtime friend who had earlier put me on the path to inventing SKYY. Over the years of our friendship, he's seen me invent and license a number of medical devices. Many of these had sprung from casual conversations with doctor friends during which they described how they performed a procedure. As you can imagine, I always asked a lot of questions and several times went on to invent or improve an instrument.

Martin had observed the need to protect medical personnel, who often experience accidental needle sticks, or "sharps-related injuries." His description of the problem—the frequency with which people are stuck, the dangerous diseases they can contract through these sticks and the inadequacy of then-current safety devices—immediately convinced me that this was a problem worth solving.

In fact there are something like 5.6 million healthcare workers in the nation today. According to the Centers for Disease Control and Prevention, these workers sustain hundreds of thousands of needle stick injuries annually, exposing them to a variety of diseases. Early in 2000, OSHA, the Occupational Safety and Health Administration, issued a Bloodborne Pathogen Compliance Directive aimed at ensuring healthcare workers access to safety-engineered sharps devices that reduce their risk of injury and infection. Congressional hearings on the subject followed.

Operating, emergency and patient room care can often be hectic, with several things happening at once, and many people getting in one another's way. Needle sticks have always been a common occupational hazard, but with the virulence of HIV/AIDS, hepatitis C and the like, the stakes are higher than ever. An estimated twelve thousand medical workers are infected with hepatitis B and HIV annually and at least twenty-nine healthcare workers are documented to have contracted the AIDS virus through needle sticks at work (*San Francisco Chronicle,* April 14, 1998).

Unfortunately, Martin's needle design was inadequate. It was somewhat complicated, using springs and tethers and requiring a longer than normal needle. He had shown his device to Becton, Dickinson and Company (BD), the world's largest manufacturer of medical devices and a pioneer in the development of healthcare safety products. They had rejected it, so Martin asked me to consider the problem.

Existing safety needles required the use of two hands. We talked to nurses, who experience 46 percent of all sharps injuries, and they told us this was impractical in many situations. They wanted a needle that could be "safed" quickly and with one hand, preferably with one finger.

We also talked to doctors, hospital administrators and purchasing agents. Final decisions are often made by the purchasing department; if it's too expensive, they don't buy it. But it seemed likely to us at the time that legislation was coming that would force the purchase of safety needles, so while we always seek the best way to solve a problem at the lowest cost, we concentrated on the needs of the end user, the nurses. We knew we had conducted enough product research with them when we began to hear the same requests over and over. When you are no longer hearing anything new—but not before then—you know it's time to move on to satisfying those needs.

The best inventions are so simple that in retrospect they seem self-evident. Al Kolvites, designer Robert Cohn and I perfected a push rod and a cover that locks. After the needle is used, it can be propelled through a hole and into its cap with one finger. The needle is locked in the cap, unable to move back through the hole because the cap is cocked at a slight angle and the needle is no longer aligned with the hole. The locking mechanism clicks into place, letting personnel feel and hear that the needle is safe without even taking their eyes off the patient. Once locked, the hypodermic can be dropped and still not perforate its cap.

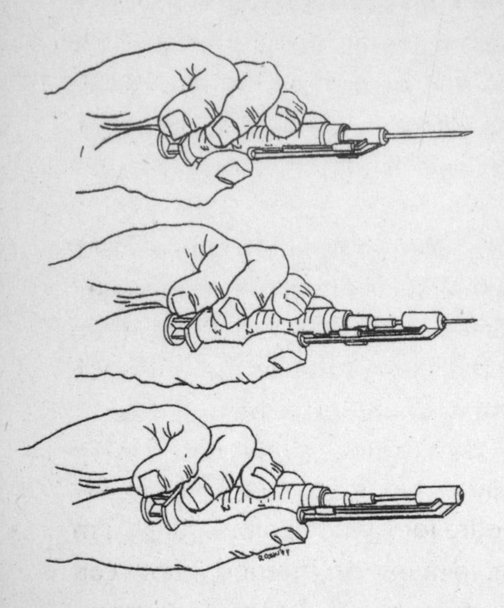

We worked with plastic and Scotch tape first and then cut bits of plastic with a razor knife and melted parts together. Reassured that we were on the right track, we did some precision drawings that further solidified our thinking. When we'd achieved our single-hand push mechanism with an easy bar-and-lock device, we applied for a patent (with an artist's renderings) and invested $10,000 in a one-cavity mold to produce a working model.

We brought the mold to BD and they were impressed. Having a mold that can produce samples, rather than only drawings which then take a large company months to bureaucratically move through the development stage, always impresses a potential licensee. Because BD is the world's biggest and best manufacturer of medical devices, working with them was our goal. We would never have tried to manufacture and sell a needle protector on our own. BD has the complex machinery needed to mass produce the needles and a broad, established distribution network. If BD had passed, we would have moved on, showing the product to other medical technology firms.

We (the patent is in the names of all four of us since we each played a creative role in its design) licensed the rights to the patent to BD for an up-front sum and royalties. My top-notch patent attorney, Mike Ebert, also negotiated a minimum yearly payment. This gave BD incentive not to sit on the patent but to produce and distribute the needles quickly. Our agreement also allowed us to take the needle elsewhere if, after a certain period of time, BD had not begun to manufacture and sell them. BD is doing so now, calling our needle the SafetyGlide Shielding Injection Needle.

A magazine interviewer once asked me, "What would you invent if you could invent anything in the world?" I answered, "A cure for AIDS." Unfortunately, I'm not equipped to turn that idea into an invention. But I can hope that our needle protector will prevent some transmissions. Solving problems always feels good, as does making an honest buck, but inventing this solution felt especially good.

This Won't Hurt a Bit: Understanding Patents

The Constitution gives Congress the power "to Promote the Progress of Science and Useful Arts, by securing for limited Times to Authors and Inventors the exclusive Right to their respective Writings and Discoveries" (Article 1, section 8). In exchange for explaining their work and putting it in writing for all to see, inventors get exclusive rights for twenty years from the filing date of the application. (Patents issued before 1995 granted seventeen years of protection from the date the patent was issued.) Patents were intended to spur innovation by creating a repository of how-to information. While one cannot make, use or sell a patented invention without risking being sued by the patent owner, one can study the technology behind the patent and perhaps improve upon it. For instance, in June of 1941, actress Hedy Lamarr and composer George

Some Names and Dates
(There will be *no* quiz!)

- *1421—History's first recorded patent for an industrial invention is granted to Filippo Brunelleschi, in Florence, Italy.*
- *1641—The first American colonial patent is issued, in Massachusetts, for a salt-making process.*
- *1790—The first U.S. patent is issued, for a new method of making pearl ash and potash for fertilizer.*

Antheil received a patent for a radio-controlled torpedo "Secret Communications System." Though it wasn't used during WW II, after the patent had expired, Sylvania modified the system and applied it to satellite technology.

Only three patents were issued in 1790, the first year they were issued in the United States. The examiners were the Secretary of State, the Secretary of War and the Attorney General. In 1999, more than 169,000 were issued, and more than 6 million patents have been issued to people around the world. Today's somewhat less titled examiners are specialists in various aspects of technology, charged with being well versed in the "prior art," or prior developments and concepts, in their area of expertise.

Examiners review a patent application with respect to three main criteria: The invention must be useful, new and nonobvious. If your invention is a problem solver, chances are it's useful. A search of patents in existence will give you a good sense if it's new, or in Patent and Trademark Office terms, "novel." Nonobviousness is trickier. The rule is that if someone else in the field would naturally have stumbled upon your innovation, it's obvious and not patentable.

The reason I've emphasized that inventions are things, not ideas, is the fact that you can't patent an idea. While you don't have to provide the Patent and Trademark Office (PTO) with a working model, you do have to turn your idea into an invention ("reduce it to practice") and provide a detailed description of the invention sufficient to make the examiner believe your invention works. If I had told the PTO that I had seen the need for a needle protector and wanted to protect my brilliant insight, they would have laughed at me. I had to provide them with detailed drawings that showed how my needle protector worked.

►*Defining Our Terms*

Patents, like trademarks, copyrights and trade secrets, are classes of intellectual property. But as we'll explore in chapter 4, patents are essentially articles of personal property that can be sold outright or licensed in return for royalties, as I did with the needle protector.

Patent. A patent is a document issued by the federal government that grants the holder(s) the right to exclude others from making, using or selling the invention in the United States for twenty years. (Under special circumstances, patents on drugs, food additives and medical devices may be extended for an extra five years.) Violating patent rights is known as *infringement* and can be litigated. (Patents can't be "renewed." After twenty years, others may copy your invention without infringing, but they cannot repatent the invention.) There are three main types of patents: utility, design and plant.

Utility Patent. This is the most common type of patent. It applies to inventions that function in a unique manner

The Patent and Trademark Office Website (www.uspto.gov) provides a comprehensive overview of patents, trademarks and copyrights, and includes detailed instructions on filing and registering.

and produce a useful result. In addition to being useful, new and nonobvious, an invention must fit into at least one of five statutory categories:

1. Compositions of matter—such as chemicals, drugs, plastics and fuels.
2. Manufactures or articles of manufacture—relatively simple (without working or moving parts as primary features) items that have been made by human hands or by machines; this can apply to everything from paper clips to buildings.
3. Machines or apparatuses—generally devices with moving parts that are used to perform a task, such as a cigarette lighter, VCR or gas engine.
4. Processes or methods—ways of doing or making things with one or more steps, such as pasteurization and software processes.
5. A new use or improvement of one of the above.

Distinctions between these categories can be blurry, and while an examiner must conclude that your invention fits into one or more of them, you needn't state which one(s). Items that don't fall into one of these categories and thus don't qualify for utility patents include naturally occurring items, mental processes, natural laws and printed matter.

Design Patent. Good for fourteen years, these protect the unique, purely ornamental shape or design of a manufactured item (for example, Nike shoe designs). The same item might have a utility patent covering its functional features and a design patent to protect its ornamental shape or design.

Plant Patent. These protect unique plant varieties that are produced through grafting or cuttings (asexually reproduced), such as hybrid tea roses or Better Boy tomatoes.

Trademark (™). These are words, slogans, symbols, designs or some combination thereof that distinguish a product or service—the brand names that give products corporate identity, such as SKYY Vodka, Coca-Cola's name and typeface, and the Nike swoosh. "Trademark" sometimes refers to a "service mark," which is the name by which a service, rather than a product, is promoted. McDonald's is a service mark (their service is selling food); Big Mac is a trademarked product.

Copyright (©). This protects published and unpublished literary, dramatic, musical and dance compositions, films, photographs, paintings, sculpture, other visual works of art, and computer programs from being copied. (For instance, the words on the SKYY Vodka label are copyrighted.) Copyrights protect the expression of ideas, not the ideas themselves. Materials that were copyrighted in 1978 or later last the lifetime of the author, artist or designer plus fifty years. Copyright protection begins automatically when the work is set on paper or otherwise fixed in tangible form. You can also obtain and fill out a form (call the Copyright Office Forms Hotline, 202-707-9100), send it back with the appropriate fee, and thus register your copyright.

Trade Secret. This covers a wide spectrum of formulas, patterns, manufacturing processes, methods of doing business or technical know-how that give the holder competitive advantage and which are kept secret, such as the SKYY Vodka distillation process or Coca-Cola's recipe.

"Patent Pending." This serves notice that you have sent a patent application to the PTO. You can use the phrase only after the PTO has received your application.

Get It in Writing: Protecting Yourself

Before, in addition to or perhaps in lieu of filing for a patent, there are several do-it-yourself ways of protecting your idea: documenting the genesis and development of your idea, using confidentiality agreements, and writing an invention disclosure.

One way of insuring your full protection rights is to document every step of your creative process, from idea through execution. The PTO recognizes the "first to invent," not the "first to file." The U.S. is, in fact, the only first-to-invent rather than first-to-file country. If in the midst of designing, prototyping and testing my needle protector, another inventor had set out on the same path, proof that I was working on the device first could have been the only way to secure my patent.

If you are serious about inventing and patenting, you should get into the habit of writing everything down. Get a bound book (something that makes the deceptive insertion of pages less likely) with numbered pages (or number them yourself). Check stationery or office supply stores for lab, engineering or accounting notebooks. Write in ink and don't tear out pages, black anything out, or leave large blank spaces. Every time you have an idea that you think has any value, make a note of it, or draw a picture of it and have one or two other people, preferably unrelated to you, understand, witness, sign, and date it. (A legend above their signatures such as "read and understood by" is appropriate.) You'd better believe that Al recorded the notes from our initial Kanbar Target brainstorming conversation. Smart inventors, especially those who have been aced out of legal protection by not being able to produce "first to invent" documentation, have shelves filled with these books.

> *Inventor's Journals, logbooks with complete instructions for inventors, are available from the Inventions, Patents and Trademarks Company;* The Inventor's Notebook, *available from Nolo Press, includes worksheets, checklists, and sample agreements. See appendix for contact information.*

You want, if need be, to be able to establish when you had your brainstorm—the "date of original conception"—and how you have made it into something real—"reduced it to practice."

Your logs and records should include virtually all of your invention-related activities: phone calls, conversations, sketches and drawings, letters, and prototype stages. If you're not sure if it's invention related or if it could ever be meaningful as proof, write it down anyway. This record keeping doesn't take much time. Do it routinely at the end of the day. Sign, date, and have witnessed anything and everything you might one day use. Writing everything down not only leaves a paper trail that may come in handy in a patent dispute, it also helps you clarify your thoughts and reminds you of forgotten details and insights.

When you show your idea before its launch or patenting, consider using confidentiality or nondisclosure agreements. (See the sample agreement in the appendix. They can also be obtained in legal form books or from your attorney.) They are a simple, important part of your paper trail and don't imply paranoia on your part or distrust of the "disclosee." They can be used when you show your developing product to people you hire to help you at various steps, to potential licensees or manufacturing partners, and even to people you ask to sign and witness your invention disclosure. Remember to give the disclosee a completed copy of the document. You can also ask disclosees to sign and date your inventor's logbook or journal under a legend stating that they have seen and understood your confidential invention, as described on such and such notebook page(s).

I am currently developing a bread product called a Wagel. I've registered the name but when I go to a baker, this product's version of a manufacturer, to help me create samples, I share my recipe, or "design." I use a nondisclosure to prevent him from using or talking about my recipe.

If someone balks at signing a nondisclosure, you have to weigh the situation and decide whether to trust the person. If you really want what they have to offer, you might have to take a deep breath and go for it, knowing that you have your first-to-invent records in order. Your recourse, if someone has signed a nondisclosure and then breached it, is suing, which is costly and time-consuming. So strive to work with trustworthy people in any event.

Once you have a solid sense of how your invention will work, writing an invention disclosure is a good way of proving its date of conception. Briefly summarize your invention, including a descriptive title of the invention (not necessarily its ultimate trademark), the background of the invention (that is, products that are currently in use), a short description of the invention itself, sketches, a detailed rundown of how the invention works and a list of its unique features. Include your name, address and telephone number, and then date and sign it. To give the disclosure added weight, either have two uninterested, unrelated witnesses sign and date it, have it notarized or fax it to your patent attorney. Keep it in a safe place. You can also register your disclosure with the PTO for a nominal fee under their Disclosure Document Program. (Call the PTO or visit their Website for complete instructions.) This is rarely advisable because once you register, you will have only two years to act on your disclosure. And remember, an invention disclosure is not a substitute for continuing to maintain full and accurate logs of your inventing process.

➤ *Conducting a Patent Search*

Whether or not you decide to apply for a patent for your invention, doing a patent search early on in your creative process is important—if only to make sure you aren't infringing on someone else's patent and opening yourself up to a lawsuit. But most of the time, you want to find out if your invention is patentable or not; if you discover it's not, you thus save yourself lots of development time and the cost of a patent application. You can do this initial search yourself or hire patent searchers, a patent agent, or a patent attorney to do it for you. Al and I recommend that you first conduct a search on your own for several reasons: doing a search can educate you about how to write your own application, about "prior art" components and technologies and about what has succeeded or failed in the past. It's also cheaper.

However, we also recommend that once you have conducted a preliminary search on your own and found no impediments, you should contact a patent attorney (or other professional) to conduct a more thorough—and costly—search. While I don't believe in wasting money, I also don't believe in cutting corners. If Al and I turn up a patent roadblock on our own, we save ourselves the money and don't bother contacting our attorney, but we also know better than to get deep into product development without having an expert do a comprehensive patent search. Cutting corners here has a way of backfiring. How would you like to conduct or pay for an incomplete, bargain-priced patent search, then spend a few thousand dollars prototyping and maybe even pitching your invention, and then find you are infringing on an existing patent? Don't let that happen. Your preliminary patent search is only a yellow light; proceed with caution until a patent attorney tells you the way is clear, then go for it. (We'll talk more about the pros and cons of hiring patent attorneys below.)

> *A "poor man's patent search" consists of checking to see if your product is already on sale in stores, through catalogs or at trade shows.*

One way to start your patent search is by going to a Patent and Trademark Depository Library. These contain microfiche copies of patent office records and often hold periodic classes on how to do your own search. Your public library may be a patent depository library, or your local university may have a branch. For the one nearest you, call 800-PTO-9199.

Internet searches, which rely on matching key words, are slightly less effective—a patent much like yours may simply use different descriptive terms. You can access the PTO online and do a search, or, if you are inclined to make a trek to Arlington, Virginia, you can go to the PTO Search Room, the most complete repository of patent information. Other patent-related sites that offer full texts of patents exist online, including Micropatent (www.micropatent.com) and IBM (www.patents.ibm.com).

How do you conduct a patent search for an invention that is so unique you have no idea what others might have called it? You need a classification number. (The PTO uses about 500 subject classes and 200,000 subclasses.) Patent and Trademark Depository libraries offer help in book and CD-ROM form, or you can write or fax the PTO, describing your idea, how it works and so on, and including a rough sketch. They'll get back to you in two to six weeks with a classification number. This is a free service: write to Branch Chief, Patent Search Room, U.S. Patent and Trademark Office, Crystal Plaza 3, Room 1A01, Washington, DC 20231; or fax 703-305-5491.

When and if you decide to hire professional help, you can find patent searchers, patent agents and patent attorneys in the Yellow Pages. In addition, agents and attorneys are listed in *Attorneys and Agents Registered to Practice by the U.S. Patent and Trademark Office,* a page-turner available in most library reference departments.

►*Applying for a Patent*

Step-by-step instructions for filling out a patent application can be obtained from the PTO (or on their Website). In simplest terms, applying for a patent goes like this: You (or you and your attorney) prepare an application and send it to the PTO. At the PTO, a patent examiner reviews it. More often than not, you will be asked to respond to objections and make changes, additions or deletions. If the application is rejected outright, you have the right to try and convince the examiner that he or she is in error. The whole process generally takes one to three years.

Your application consists of a self-addressed receipt postcard, a check for the filing fee, a transmittal letter, drawings of the invention, the "claims" you are making about your invention, an "abstract" summary, the Patent Application Declaration Form and the "specification," which describes the invention. There are several parts to the specification, ranging from the title of the invention and its background in terms of prior art to a description of its operation and a discussion of alternative embodiments. (See why I have a patent attorney? Just thinking about this makes me want a SKYY martini.)

The rights granted by a U.S. patent have no effect in foreign countries. If you want patent protection in other countries, you must apply in those other countries. Since the laws of other countries vary, working with a patent attorney skilled in obtaining foreign patents is advised.

At this point, you might be wondering: Why bother with all of this? Simply put, patent protection preserves your right to make money with your innovations. It's hard to command a price from a licensee without a patent, and having a patent can sometimes help you to raise money. Also, having a patent can slow down your competitors, who may have to redesign their similar product in order to avoid infringing on your patent. However, while it's true that inventors in the United States enjoy a very high level of patent protection, and the courts usually find in favor of patent holders in patent infringement cases, some people don't believe the benefits outweigh the costs. Patents are expensive to apply for in the first place, and defending a patent can be very costly and time consuming. It's worth reiterating that having a patent is no guarantee that you'll be able to manufacture, sell or license your invention. Many, many existing patents are not protecting existing products. Maybe the patent holder discovered there was no market for the invention or perhaps a better product was produced and patented in the same field before he or she could get to market. Some people get patents simply so that they can frame them and hang them on their walls. This is silly.

Let me also reiterate that protecting your idea isn't the same thing as evaluating your idea's viability in the marketplace. You must do both if you are serious about marketing your innovations. Assessing and ensuring your ability to get a patent is part of exploring the viability of your invention.

It is also important to remember that obtaining a patent on your invention does not give you the right to "practice" that invention, only the right to exclude others from making, using or selling it. The PTO has no jurisdiction over questions relating to infringement of patents. In examining applications for patents, they make no determination as to whether the invention infringes any prior patent. An improvement invention may be patentable, but still might infringe a prior unexpired patent for the invention improved upon. Even though you have an issued patent, your invention may still infringe someone else's patent.

Deciding when to apply for a patent depends partly on the specifics of your invention and your plans for manufacturing or licensing. Here are some things to consider:

- Applying before you assess the needs of the marketplace and before researching existing patents is never advisable.
- It's ideal to have a patent or patent pending status before approaching a potential licensee.
- Getting a patent before you build a prototype can be risky. You may find you need to modify your plans significantly enough in your prototyping process to make the patent protection you've applied for inadequate.
- If you know your invention is a fad or fashion-sensitive item that will realistically be marketed for a limited amount of time, or if your estimated profits will be fairly small, you should consider skipping the time and expense of patenting. Unpatented or unpatentable ideas can still produce profit.

WARNING:

"Inventors are reminded that any public use or sale in the United States or publication of the invention anywhere in the world more than one year prior to the filing of a patent application on the invention will prohibit the granting of a U.S. patent on it. Foreign patent laws in this regard may be much more restrictive than U.S. laws."

—U.S. Patent and Trademark Office Website (www.uspto.gov)

One patenting strategy is to put your product on the market before patenting it. This way, you can put the money you would have spent on patenting into marketing and hopefully generate some income before undertaking the expense of a patent. If you find your product is a dud, you can skip patenting altogether. There are risks with this approach. You can file for a patent up to one year after you first sell or otherwise disclose your invention, but by taking advantage of this delay, you run the risk of someone else filing first. A better strategy may be to file for a provisional patent just before product launch or disclosure. Laws that went into affect in June 1995 allow independent inventors and small companies to file a "provisional patent application." Less costly and intricate than a regular patent application, the provisional still allows you to post "patent pending" on your product for a year while making sales or securing a licensee. Should your results be disappointing, you needn't file a more expensive regular application. Doing this late in your development process allows you to include the most accurate and broad terminology in your patent and secures you some protection from competitors. After filing a provisional, you have one year to file a regular application. Your provisional and regular applications should not contain significant differences, which is another argument for filing the provisional late in your development process.

I've never done this, but another strategy is to apply for a patent yourself (and save the attorney fees) when you know your product will not be eligible for significant protection. By applying yourself, you still enter the "patent pending" phase, you may delay competition, and you may improve your chances with investors and licensees. You can let your application expire if you decide not to proceed before the patent is issued.

An argument in favor of patents is that they protect you when you pitch your invention to potential licensees or manufacturers. (We'll talk more about this in the next chapter.) Idea theft is not rampant, but it does happen, so don't risk it. If you don't want to wait until you have full patent protection, which as we've seen can be costly and time-consuming, "patent pending" status can deter thieves. This is your status after filing either a provisional or a permanent patent application. If you file neither of these, use a nondisclosure agreement. Legally, once you've told your trade secret, you can no longer claim exclusive rights to it.

> Periodic maintenance fees must be paid to the PTO in order to keep an issued patent in effect.

If you're beginning to feel that figuring out when, how and why to apply for a patent is too complicated to decide on your own, keep reading.

►*Hiring a Patent Attorney*

You don't need to employ a patent attorney in order to obtain a patent. But there are several good reasons to consider hiring one, not the least of which is that he or she can help you develop the best strategy for your situation. In my opinion, you should only write your own patent application if, first, you are already a talented writer, and second, you are motivated to take time away from inventing to learn the ropes of the process. A good patent attorney, on the other hand, can

1. probably write a better application than you,
2. do a more thorough patent search than you,
3. save you time, and ultimately
4. save you money.

Inventor's groups, such as the United Inventors Association of the USA (716-359-9310), may provide tips on good patent attorneys near you.

My patent attorney for more than thirty-five years was Mike Ebert. He's a wonderful man with an extraordinary ability to draft strong patents. When he was a young attorney just starting out, the higher-ups at his firm handed him a task they didn't want—writing a patent for a newfangled copier. This Xerox patent is now featured in law textbooks as a model of solid patent writing. Mike's work was so good that Xerox tried, and failed, to talk him into going to work for them. Their loss was my gain.

Mike has recently retired and we miss him. Our long relationship taught me what to seek in a replacement. Most basically, you need to be able to communicate with your attorney. I can call Mike on the phone, describe my idea and detail its mechanics, and Mike will "get it" instantly and start writing it up. If an attorney has a different understanding of your invention, or if he or she doesn't quickly get your drift, go elsewhere.

Some people advise shopping for a patent attorney who has experience as a litigator. Their reasoning is that these attorneys may be more careful about avoiding loopholes in their patent writing. You might also consider looking for a patent attorney who has written patents for other inventions in your technical field, or seek out someone who has an educational background in your invention area.

Once you've finished your own preliminary patent search, and you're interviewing potential patent attorneys to write the actual application, here are some smart questions to ask:

- Is my product eligible for a broad claim?
- Will I be able to obtain a single-feature claim?
- To what degree will a competitor have to change my product to avoid my patent?
- Which of my product features will a patent protect?
- Do any of the patents I've uncovered prevent me from selling my product?

- If I do get a patent, are you (the attorney) willing to take a patent infringement case on a contingency basis? (They can be extremely costly.)
- What parts of the product can't be changed without losing protection, once I've applied for the patent?*

If a patent attorney is convinced your invention will be a moneymaker, he or she may agree to provide services against future revenues.

While you must be able to trust your patent attorney and the other experts you work with, you can't turn off your own brain. Educate yourself so that you can make informed decisions with your expert advisors. Trust your gut and be willing to take calculated risks, not because you're a daredevil who's hell-bent on ignoring all received wisdom, but because you have done your homework and have the courage of your convictions. As James Gleick recently wrote in the *New York Times Magazine*, inventors are "a familiar species of fraud victim. An entire industry of invention promoters promises to help inventors get patents, usually charging thousands of dollars in fees that are virtually never recouped" (Gleick 12 March 2000).

You avoid becoming a fraud victim by becoming savvy. My D-Fuzz-It patent attorney cautioned me, warning that getting a patent wouldn't guarantee me a damn

---

* *Entrepreneur Magazine: Bringing Your Product to Market*, Don Debelak. Copyright © 1997 by Don Debelak. Reprinted by permission of John Wiley & Sons, Inc.

thing. I took his real-world warnings to heart and forged ahead anyway, certain that I could market the device. I'm glad I did. Conversely, when my friend Knud Dyby was advised by a model maker not to seek a patent for his ingenious paper clip holder/dispenser, he should have gotten a second opinion from a good patent attorney. We seek advisors who are knowledgeable and trustworthy. Cultivate those qualities in yourself so that you can exercise good judgment about taking and overriding any advice you receive.

A patent attorney can obtain a thorough patent search and then assess its results for you. Unlike your preliminary search, this will include a complete domestic and worldwide search conducted at the Patent and Trademark Office by an experienced searcher who uses both manual and computer search techniques. This person will confirm with a patent examiner that the search is conducted in the proper fields. The patent attorney will then review the search and give a legal opinion on your potential patent's commercial value. This is another reason to be wary of inventor service organizations, which benefit by telling you that you have a great idea and should file immediately. They'll get more money out of you that way, whether it's true or not. Use an attorney you trust instead.

A patent attorney can save you time because, as you've probably gathered, assembling a solid patent application is time consuming. (Wouldn't you rather be inventing?) What's more, it takes skill. There are solid patents with "teeth" and "loose" patents that don't provide much protection.

The claims are the most important part of the application. They determine the scope of your patent rights. A skilled attorney will write your claims in a way that makes them unassailable. A strong patent is one that makes a straightforward, simple claim with one or two key words or phrases. The more key words or phrases, the weaker the patent, since another invention has to infringe on them all to warrant your protest. Patents with broad claims are strong because a wide variety of specific designs may infringe on that patent.

I couldn't have patented the entire "needle protector" concept, just my needle protector. But I could describe my needle and make claims about it that ward off infringement. For instance, if I had defined a particular material as being integral to my device, another inventor who used a different material could say her device is different. In our first claim for the needle protector, we describe "a push rod of flexible material slidable on the track and terminating at its lower end on a finger rest which when engaged by a finger of an operator advanced the rod from a retracted to an extended position." The operation is described with no limiting details. The finger rest shape is not detailed and the "flexible material" is not defined.

Finally, despite their up-front cost, patent attorneys can ultimately save you money because missed deadlines and other filing errors can cost hundreds of dollars in extra fees. Plus, getting the fullest patent protection possible could well save you money in the long run by protecting you against infringement litigation.

chapter four

MANUFACTURE OR LICENSE?

It's a Puzzle: Tangoes

Some twenty years ago, at the home of friends, I watched a child as she attempted to assemble a seven-piece puzzle. She was completely engrossed by the effort, and I asked her mother about the puzzle, which was unlike any I'd ever seen.

My friend explained that it was a Tangram, an ancient Chinese puzzle made up of seven pieces, or tans, of various shapes. The goal is to assemble the pieces into various forms: animals, objects, geometric designs. It seems easy but putting the pieces together correctly can be a real challenge.

My friend said the puzzle originated hundreds of years ago when a man named Tan dropped a porcelain tile. It broke into seven pieces, and he had a devil of a time fitting them back into their original square shape. As Tan kept reassembling and reordering the pieces, various figures emerged—birds, people, buildings and so on. Legend has it that word of the brainteaser spread, and it became a favorite pastime called Chih-hui-pan, or the Wisdom Board.

In the early 1800s, a book of the puzzles appeared in China. The puzzle then spread to Europe, where Napoleon reputedly passed his time working with it on Elba. Lewis Carroll is said to have been so devoted to Tangrams that he carried a book of them with him at all times. In the early 1900s, Sam Loyd wrote *The Eighth Book of Tan,* in which he put forth the claim that Tangrams were invented by the god Tan more than four thousand years ago. The puzzle is rumored to have been brought to America by Western sailors who stopped in China during the opium trade. To this day Tangram puzzle problems appear in Chinese newspapers, much as crosswords appear in Western papers. In modern classrooms, math teachers use Tangrams to teach spatial relations, and puzzle devotees can find dozens of versions of the Tangram in books or on Websites.

I've also heard a story that traces the Tangram's origins to a hapless window maker employed by a king. When the artisan dropped the square window he had been commissioned to craft, it broke into seven pieces. As he tried to mend it, a winged horse emerged from the broken pieces. Each new attempt to mend the window elicited a new creature. Happily, the king loved his magic window. For me, I'll stick with Tan and his porcelain tile, but clearly the story of the square broken into seven difficult to realign pieces has enduring resonance.

Back at my apartment after the evening with my friends, I began to wonder how I might make this intriguing puzzle better. Figuring out how to improve an existing product or technology is the same kind of catnip for inventors as the problem-solving challenge. Observation and curiosity often lead me down this path, which can be lucrative if you can make a product more appealing to its current market or open up additional markets. (SKYY Vodka has become popular not only with vodka drinkers but also with former non–vodka drinkers attracted by its "clear" benefits.) Whether I'm brainstorming my next innovation or just indulging in the inventor's version of counting sheep, I often put myself to sleep at night by letting my mind survey how-can-I-make-it-better strategies.

As I lay awake that night, it seemed clear to me that, whether he was a god or just clumsy, Tan's invention had lasting appeal. Could I improve on it? I thought about turning it into a game, with speed and competition added to the goal of simply completing the puzzle. Two players could challenge each other or a single player could try to better his or her own best time. Unlike a chessboard or Monopoly game, I thought the Tangram game should be small and portable, like the Rubik's Cube. I saw the appeal of something that could be pulled out of a pocket, backpack, briefcase or glove compartment, and either fiddled with for a few minutes at a time or played with for hours by an entire family after dinner.

In a flash, I came up with what I thought was a great name. Tangoes refers back to Tan and Tangram while also conveying the back-and-forth nature of a game for two players or teams.

The real challenge of making my idea real was in creating its form. I wanted a compact box that contained all the pieces, puzzle shapes, and solutions. I wanted the box to be part of the game, something unique that added to the appeal rather than just a container for all of its parts.

I came up with something so unique it turned out to be patentable. I obtained a utility patent covering the shape of the box and the way it functions as an easel. The sleek plastic Tangoes box is only 6½ inches long, 3¼ inches wide and just over ½ inch thick. The cover slides off to reveal two sets of identical puzzle pieces (triangles, squares and parallelograms) and a pack of cards featuring fifty-four different puzzles on one side and their solutions on the reverse.

The players slide off the cover, take an identical set of seven puzzle pieces each, shuffle the cards, and then replace the cover so that cards can be slid out of the open end. They then slide out the first card and place it in the slot on top of the box cover, which acts as a kind of easel. The first player to solve the puzzle, using all seven pieces, announces, "Tangoes!" If the player has solved the puzzle correctly, he or she keeps the card; if not, the opponent gets it. The first player to gain five cards, or any agreed upon number, wins.

Players are urged not to be limited by the designs provided. If while trying to make one of the given figures, they find they've designed a new figure, they can simply slip a piece of paper beneath the tans and trace the outline. Removing pieces one at a time and filling in the lines gives them a permanent record of the solution to their own design. Players can also consider challenging themselves and others to create new figures.

Both kids and adults have found this game fascinating. Working the puzzles quickly requires visual and spatial problem-solving skills. Players learn to think creatively, artistically and even mathematically.

I took my game to major game and toy manufacturers—Parker Brothers, Ideal Toys, and Hasbro, Inc. Parker Brothers showed real interest and we had several discussions. They ultimately passed on it, saying that their research showed they would only sell a hundred thousand units a year.

I thought about that. A hundred thousand units a year, huh? If I made $2 a game after all costs, that was $200,000 gross. If it cost me $50,000 to run the business, that still left me with $150,000 take home, which ain't, as they say, cornflakes. I decided to give it a try.

As I've already said, around this time I met a young man named Mark Chester. He was a great salesman and since I knew I wanted to use a contract manufacturer to produce Tangoes but didn't want to handle its marketing, we made a deal. I offered Mark a piece of the pie (or should I say puzzle?) and he ran with it.

My kick is inventing. I let my D-Fuzz-It partner buy me out after a few years, have given over the day-to-day running of the Quad to my brother, Elliott, and have been happy to license my numerous medical devices. I could arguably have done better with some of my inventions if I'd held the reins longer and more tightly, but even with SKYY, the invention I've marketed most personally, I have come to turn over more and more of the daily operations so that I could move on to more inventing.

Mark and I formed Rex Games around Tangoes in the early 1990s and had the game manufactured in Taiwan (and now in China). Tangoes did terrifically well from the start—much better than Parker Brothers' estimates. A couple of years later we were grossing nearly $2 million a year. Earning placement in the prestigious New York Museum of Modern Art gift shop and catalog was a gold mine as they influence museum shops and catalogs around the country.

Like the D-Fuzz-It, Tangoes has been a perennial moneymaker. At Rex, we strive to produce games that build on the Tangoes model, teaching creative, strategic and logical thinking. The games foster math, language and reasoning skills in players of all ages, while teaching the young about sharing, winning and losing, and taking turns—they teach the mind to think and no batteries are required; our games work on brain cells.

Making It: When to Manufacture

You've diligently perfected your invention, and perhaps applied for a patent. Now what? Most inventors, including this one, would be ready to make some money. Even if you have decided not to pursue a patent, you have something to make or sell.

In and of themselves, inventions and patents don't produce income. You make money with your patent and/or product by manufacturing or licensing. Actually, it may be more accurate to say that your decision is whether to go into business yourself or to form an alliance with another entity. There are many options within each choice.

If you go into business yourself, you might undertake everything—manufacturing, distribution and sales—or you might contract out one or more of these tasks. Indeed, I don't own or operate any manufacturing plants; I contract with various manufacturers to produce my products to my specifications.

If you don't launch your own product-based business, you may sell the rights to your product outright or license them to some degree. By entering into a licensing agreement, you (the licensor) grant a company or individual (the licensee) the rights a patent gives you—to manufacture, sell and use your invention. You can also negotiate sales- or manufacturing-only licenses. Another option, put forth by inventor and marketer Bob DeMatteis, is creating a job for yourself within another company. As an in-house product development manager, you can develop your patent as an employee who receives a salary and stock options instead of royalties (DeMatteis 1997, p. 19). The possibilities are truly limited only by your creativity, powers of persuasion and personal goals.

It's difficult to sell or license your invention unless it is patent pending or patented, though in rare cases you might be able to sell or license it as a trade secret, or as some other form of intellectual property. Most business deals involve an exchange—I'll give you this and you'll give me that. The rights granted to you by a patent give you something tangible to sell or license, and they prove that the invention is yours to sell. That said, though Knud Dyby, the paper clip dispenser inventor, didn't obtain a patent, he did profit from his invention. The injection molder who advised him against getting a patent did recognize the value of the invention. He manufactured the product and shared a percentage of the profits with the inventor.

There are a number of advantages to running your own business and/or overseeing your own manufacturing: you control quality, you maximize your profits (although these can also be substantial with savvy licensing) and you can pursue product improvements and additions as you see fit. Of course, starting a company requires the necessary business skills, lots of time, and a stomach for risk taking, not to mention the need for start-up capital (see chapter 2).

Potential problems to bear in mind are that a company that manufactures and sells only one product is sometimes a low priority for prompt payment from distributors, wholesalers and retailers. Also, it can be hard to sell just one product in a field; this continues to hinder the success of the Kanbar Target and ROLLOcane. Dentists and medical personnel we approach ask to see our catalog, expecting us to offer a range of dental or medical devices. We don't. If our sales force represented a variety of products, we might be better able to offer promotional incentives.

I decided to manufacture and sell Tangoes because we learned that we could go into stores and successfully sell the game ourselves. Today there is an entire catalog of Rex Games products, but even with Tangoes alone, we found we could get people's attention and make sales. Partly this is because the toy industry is much more open to fledgling companies if they have what looks like a hot product. In many fields, this is not the case and licensing would be much more prudent.

Deciding which money-making route to pursue depends partly on you—your interests and skills—and partly on your product. Are you entering an industry in which it's better to take advantage of an established corporation—with existing distribution networks, research and development departments, a well-known marketing presence and financial resources—or does your product have a better chance for success and future growth on its own, guided by your unique vision?

The key to success isn't much good until one discovers the right lock to insert it in.

—*Tehyi Hsieh,*
Chinese Epigrams Inside Out

Don Debelak, who has helped dozens of inventor/entrepreneurs market their new products, believes that solid products aimed at a small market are the most feasible to manufacture and market on one's own because without a big potential such products can become "lost in the shuffle" in a big company (Debelak 1997, p. 63). I'm better off that Parker Brothers passed on Tangoes; the prospect of selling a hundred thousand games was fine by me, and taking the risk myself has led to the creation of an interesting and enjoyable company, which I hadn't foreseen at the start. In addition, as Debelak stresses, you don't need to make strict either/or manufacturing and licensing decisions. Inventors who go into business for themselves can still make all kinds of strategic alliances with others—to distribute, market, manufacture and so on—that fall outside the traditional licensing arrangements in which the inventor prototypes, patents, makes a deal and moves on.

I have let my inventions dictate their most advantageous route to the marketplace. I was confident that I could produce, distribute and market the D-Fuzz-It effectively. I understood the materials needed and educated myself about the intricacies of production. I even designed my own packaging and was confident that retailers would want to stock the product if I could get samples into their hands. I took a similar route with Tangoes and, as we'll see in chapter 5, with SKYY Vodka.

Conversely, I am not the most effective marketer for my medical devices. These products need what a Johnson & Johnson or Becton Dickinson can provide: the ability to mass produce intricate designs, a strong and reputable market presence, and established distribution and sales channels. It never crossed my mind to manufacture my needle protector on my own, and I might have done better with the Kanbar Target and ROLLOcane if I had formed a licensing arrangement with such an entity.

Another thing you must consider is the cost of production weighed against a potential retail price. You can bet that any potential licensee will be looking at this closely for themselves. How much will people pay for your product in the marketplace? I don't use any hard and fast pricing formulas. Being able to retail for approximately four times your cost, including manufacturing, packaging, insurance, returns, waste, etc., is a standard goal ($1 to make, $3.98 retail). I'm sometimes willing to work with less of a markup because I have less overhead than people who maintain large payrolls and conduct expensive ad campaigns. We'll talk more about product pricing in terms of your marketing strategy when we look at SKYY Vodka's genesis.

➤*Hiring a Manufacturer*

Even if you handle distribution and marketing yourself, you probably won't set up your own manufacturing plant. The typical route that I've followed if I haven't licensed is to research and develop a product, apply for a patent and then hire someone to make a mold. (If you can't afford the mold, offer to pay the manufacturer a higher per piece price if he'll front you the mold.) I then own the mold and can take it elsewhere if I ever become dissatisfied with the manufacturer who made it. I've sometimes contracted with another entity to assemble and package the manufactured parts (a veterans' organization that provided such work actually assembled the early D-Fuzz-Its). Then it's time to either represent and sell the product yourself, or seek out a distributor (more on distribution in our SKYY discussion as well).

Some Ways to Find a Manufacturer

- *Ask someone with experience in the industry.*
- *Go to industrial and trade shows.*
- *Read trade publications. Manufacturers will advertise.*
- *Contact local inventor associations and chambers of commerce.*
- *Ask an industrial designer for leads.*
- *Go to stores and research who manufactures similar items that are already being sold.*
- *Ask your potential retail or industrial customers for a recommendation.*
- *Look in the* Thomas Register *(in print or online, www.thomasregister.com).*
- *Log on to www.bigbook.com, similar to the* Thomas Register.
- *Let your fingers do the walking—through the Yellow Pages.*
- *Consult embassy trade development councils for manufacturers in other countries.*

Most often, inventors seek out domestic or overseas manufacturers who can meet their needs, craftsmen who are expert at making a product to a client's specifications for an agreed-upon price. Even so, you must become an expert in order to work with experts. Learn the terminology, understand the processes and know the materials.

You'll want to evaluate a manufacturer's quality and consistency. Examine other products they have produced. If you have any questions about their finances, ask to see an annual report or profit-and-loss statement. What is their reputation for service, commitment and innovation?

> If you want your eggs hatched, sit on them yourself.
>
> —*Haitian proverb*

When you interview manufacturers, ask them to sign a nondisclosure agreement and/or have a patent or patent pending in place. Get references and price quotes. Can they deliver at your target price? Especially when working with overseas manufacturers, triple-check the production and shipping time. Resist being talked out of the product quality and attributes you want. I persisted and insisted on top-quality plastic for the Tangoes case, special inks and finishes for the SKYY label and a certain type of chocolate for Vermeer. Your product should be market oriented, not production driven. When you make sales or get order commitments, draw up production contracts with your manufacturer with the help of your attorney.

If you do set up manufacturing capabilities of your own, remember to scale your production to the size of your market so that you don't tie up more funds than you need to in people, equipment or inventory.

I've talked about the patent I was granted for conserving power by using LEDs in traffic lights. I have so many business "irons in the fire" that I haven't yet had time to exploit this patent. What I will most likely do is forge an alliance with someone who has the time and skill to approach cities, explain the benefits of my invention and make deals—much as I did when I joined forces to sell Tangoes. One of the skills of an inventor/entrepreneur is cultivating your ability to work with others and to find mutually beneficial partners.

►*How to Make It Even Better*

If you build your own business around your product, you will want to stay alert for opportunities to improve that product. The goal of keeping customers happy—and purchasing—is obvious. Throughout your initial product development and testing, you made modifications to meet the market's criteria. This doesn't end when your product is selling.

Of course, listen to your customers first—they'll tell you what improvements they'd like to see. But even if sales are humming along, you'd be wise to consider product modifications that keep you ahead of your competition and that maintain or add to your product's appeal. Over the years, I've improved SKYY Vodka, achieving ever lower levels of impurities. No one told me to do so, but I knew it could be done and could only boost our reputation. Doing so upped the ante for our competitors before they even came to the table. Though I've never pursued such a strategy, some product improvements can also lead to additional patents that extend your proprietary position.

Expanding a product line—as we did with Tangoes, and as we are doing with SKYY Citrus—is another way to offer consumers something new and to continuously generate attention and interest. From a marketing perspective, product improvements and variations also increase your market presence, often quite literally by requiring more shelf space. Considering how you can make your product even better is every bit as important after you've established your product as it was during your initial innovation. But never assume something is better just because it's different—only make modifications that genuinely improve the final product.

Making a Deal: When to License

As Edwin Bobrow notes, "Thomas Alva Edison died a wealthy man. . . . [But] he did not die as wealthy as he should have. Much of his fortune was plowed into 'inventing' one new company after another and fighting legal battles associated with them. Somebody should have talked to him about licensing" (Bobrow 1997, p. 237).

Most inventors today don't need that talk. They already know about licensing, and for some, it is their only goal: they are quite happy to exchange their right to make and sell their invention for a healthy fee and a fair royalty. How do you achieve this? It's up to you to pursue the companies you hope will sign those royalty checks. First, we'll talk about making your pitch to prospective companies, and then we'll review all the aspects of a standard licensing deal and its main variations.

You can make your case to a company most effectively when your invention has been reduced to practice and you have something *real* to show them. You'll make your case most safely when your invention is patented or patent pending. When patents are not involved, remember those nondisclosure agreements.

But bear in mind that nondisclosure agreements are not ideal. If you feel yours has been violated, you must incur the expense of suing for damages. And companies are sometimes reluctant to sign them. What then? Consider disclosing "around" your secrets. Can you convey what your invention does and why it will be profitable without revealing all of its mechanisms? Having heard this much, the other party may be so curious that they become willing to sign your confidentiality agreement. But even if you proceed without a signed agreement, you can establish an implied confidential relationship that can provide you with some legal protections. (See Nolo Press's online legal encyclopedia at www.Nolo.com for more on this.) This occurs in several ways: if the entity you are giving confidential information to solicits it from you without your prompting; if you make clear that you are presenting your invention as a business proposition and are looking for payment; if you request that they keep the information secret; and/or if the information is a trade secret with commercial value that is not known to competitors.

Research a potential licensee's track record, even if they are large and established. Look for a company that is familiar with your type of product; they may have to do less retooling of their own manufacturing process to incorporate it into their business. Look for one that does high-quality work and markets efficiently. Companies that don't maintain large and active research and development departments of their own may be more open to outside inventors and inventions like yours.

You can research companies at trade shows and in trade magazines, industry directories, the *Thomas Register; Standard and Poor's Register of Corporations; Directors and Executives;* and *Dun & Bradstreet's Reference Book of Corporate Management; Million Dollar Directory;* and *Middle Market Directory.* Request literature, brochures and catalogs from companies you are considering approaching.

Once again, be wary of invention marketers who charge large fees, promise to find you the perfect sugar-

[*For a fee, the Patent and Trademark Office will publish a notice in their official gazette that your patent is available for licensing.*]

daddy licensee, and tell you not to worry about the details. What are their credentials? What is their matchmaking success rate? Do they want a big up-front fee and a hefty chunk of your royalties? Can they prove they have contacts you don't? There are legitimate invention marketers, but even if you employ one, you should educate yourself about every aspect of your invention and its potential.

Pitching your product in person is preferable to doing so through the mails. To get an appointment with a company you've determined might be a good match for your invention, call the company and, depending on its size, ask for the president, the marketing director, or the person responsible for marketing your type of product—not the engineering or research and development departments. Identify yourself and why you are calling—because you have a new brainteaser game that would fit nicely into the company's product line, for instance.

You can also approach a company by sending a short letter and requesting an appointment. Let them know you'll be following up by phone in a few days and then do so. You might also consider sending a brief product summary and pitch, but a personal presentation and a hands-on demonstration of your working model is much more effective.

You can usually get your foot in the door to make your pitch. How do they know you haven't got the next Frisbee?

I was easily able to show Tangoes to Parker, Ideal and Hasbro, even though I'd never created a game before. If you don't have a track record with the company or in the industry, you may be seen by a low-level employee, but in any industry where many of their products come from outside inventors, you can usually be seen by someone.

As I've said, when it comes time to pitch your invention, I have found that it's always best to have a working model rather than a crude mock-up or drawings. Partly this is because a "finished" product is always more impressive. But mainly it demonstrates your seriousness and professionalism, and if your invention is accompanied by a patent or patent pending, you show that you've done your legwork and aren't a daydreamer who thinks someone else can turn your idea into an invention with a little time, money and effort. Part of your appeal is that you have done the research and development work—you've done the hard part. Big established companies tend to be slow, lumbering and bureaucratic about introducing new products. If you have something that works, that they need in their product line and that you've developed to a working state, you truly have something to offer them.

Your presentation or pitch should do several things:

- Show that your product has a sizable audience. Be specific about your target market and its size. Realistically estimate yearly sales and profit. Detail the product and market research you have done (outline your product's Ten Commandments) and the ways you have modified and improved your product based on testing feedback.

- Prove that your product is marketable. A good name can be helpful in this regard. Have you sold the product in a limited test market? If you are seeking a manufacturing partner, can you show that you have distributors waiting in the wings or retailers ready to buy?
- Explain how the product fits into the company's overall goals—which you've researched before pitching. Tell them why you have chosen them as a potential partner. Becton Dickinson, in addition to being our first choice for the needle protector for other reasons, wants to maintain their reputation for offering pioneering safety devices. We knew our product matched their industry profile.
- Demonstrate that the product will be easy to distribute.
- Convince them that your product is better than what they've come up with themselves. BD was already selling safety needles; ours was better.
- Anticipate any problems your product might encounter and show how you've designed around them.

If there is a compelling, amusing, or otherwise interesting story behind the development of your idea, briefly tell it. Keep your pitch short. Use visual aids if appropriate—your model, most importantly. Show that you've done your homework, but don't bore listeners with every detail. You can leave behind attractively arranged and presented written information instead, including any documentation that backs up your assertions.

If your product is rejected, politely ask why so that you can do better with your next pitch.

➤ *What's the Deal?*

If your talks with a company proceed, the conversation will come around to what sort of deal you're looking for. There are two basic options: you can sell your patent outright for an agreed-upon price with no future royalties (which I have never done), or you can strike a licensing deal whereby you retain ownership of your patent while allowing the company to make, use or sell the invention in exchange for royalty payments for an agreed-upon period of time. It is at this point that you'd do well to involve your attorney, especially if he or she is a savvy negotiator. The deal you are being offered may contain provisions that are not in your best interests and which you may be able to negotiate out of the deal. In order to do that, you'll need to understand any legalese in the proposed deal. I can't think of a situation when the goal of getting the best possible deal for yourself wouldn't be furthered by a consultation with your trusted patent attorney.

Negotiated royalties for licensing deals generally fall between 1 and 10 percent of the product's factory price—the money received by the manufacturer when it sells your invention-based product. They vary from industry to industry. Regardless of the industry, some factors that will influence your negotiations are whether you are providing an exclusive or nonexclusive license, how developed your product is, the competition it will face in the marketplace, the size of the licensing territory and so on. You can negotiate a staggered royalty that will increase after x number of units have been sold. Royalty rates may also take the form of a flat rate, say $1 a unit.

If you can, negotiate a minimum annual royalty, which means that the last of the year's quarterly payments must add up to at least the agreed-upon minimum. These are desirable for many reasons, including the fact that they encourage the licensee to manufacture and sell your product, since they have to pay you your minimum even if they don't sell one widget. My needle protector licensing arrangement made this provision. Also attempt to negotiate that any up-front sum you receive for signing an agreement not be termed an advance against future royalties. It should instead represent payment for the work you've already done.

And give yourself an exit strategy. Our needle protector license specified that if BD didn't produce the product within a specified amount of time, the patent would revert to us and we could take it elsewhere.

> *"Do not, under any circumstances, assign [transfer ownership of] your patent in return for a series of payments: if your assignee defaults on the payments, you'll be left without your patent or your money, but with a big legal headache—getting your patent back. If someone wants to buy your patent for a series of payments, see a lawyer or legal forms book and make a suitable license with an agreement to assign only after all payments have been made."—David Pressman, patent attorney and author,* Patent It Yourself *(Pressman 1997, ch. 16, p. 14)*

It's also possible and preferable to specify that if the licensor and licensee cannot resolve any dispute that arises under the license, mediation and/or arbitration will be sought and be considered final and binding. Court litigation is expensive and independent inventors generally do not have resources comparable to those of their licensees. I use binding arbitration agreements as a matter of course.

It is not unheard of for a company to enter into a licensing agreement with an inventor in order to eliminate his or her invention as competition. If you make a royalty-only deal, and the licensee never makes and sells the product, you don't see a dime. If a company professes to want to make a deal but resists anything but a royalty-only deal, it's possible their aim is simply to keep your product off the market.

While the amount of your royalty is an important part of your license agreement, it's only one of many issues. A well-negotiated license agreement ensures that you are paid fairly, it defines your rights and any limitations on the licensee, and it establishes what will happen as changes come up in the future. A comprehensive license agreement will cover all of the following areas:

- Whether the license is exclusive or nonexclusive. Manufacturers will prefer exclusivity and my licenses have all been exclusive. A nonexclusive deal, which means that you retain the right to also license others, is standard when very valuable technology is involved. For instance, three Bell Labs inventors invented the transistor. The inventors won the Nobel Prize; it was a momentous, far-reaching discovery.

Bell Labs licensed anyone who wanted to make it. They even taught people how to make transistors.

- How long the license will be in effect. Most licensees will want the life of the patent. While licensing for shorter periods of time may give you the right to renegotiate, you will want the license to last for as long as the product is being sold. A long-term relationship under fair financial terms is fine.
- A definition of the product to be licensed and whether its subsequent variations, derivatives and modifications are included in the agreement. It's in your best interests to obtain this broad wording, just as broad claims in your patent are in your best interests.
- The geographical and market scope of the license, such as the U.S. and kitchenwares. You may want to make these kinds of specifications if you intend to make separate deals overseas or in different industries, licensing your new fiber to different companies for use in apparel, sporting goods, etc., for example.
- Whether you are willing to give the licensee the rights or right of first refusal to future inventions. This is restrictive for inventors.
- Your right to audit the company's books. While I've never taken advantage of such a clause, no reputable company should refuse to have one in their agreement.
- Who is responsible for pursuing patent infringements. You'll likely want to avoid being held responsible for pursing infringers—it's expensive to do so.

- Any agreed-upon rights to sublicense. The licensee should need your permission to make such arrangements—you'll want to negotiate how to share the profits of any such deals.
- Provisions for the licensee's bankruptcy or company sale. You want the right to terminate the agreement in either event.

Finally, you should always have your attorney draw up your agreements, or at the very least have an attorney review them before signing.

When I developed an improved instrument for treating varicose veins, you'll recall that I invested in a mold with which to make samples. I gave samples to the doctor friend who had first spoken to me about vein-stripping procedures. He gave some samples to some of his colleagues and one way or another, Johnson & Johnson wound up calling me—not, I'm sorry to say, the usual way things work when you seek to license. Johnson & Johnson requested a meeting with me. At that initial meeting they asked if I could give them 150 samples to test market. Because I had made the mold, I could.

Johnson & Johnson got a terrific response and quickly called me back in. I had travel plans and asked if we could meet later in the month. They asked me to rearrange my plans and come in pronto. I did and we quickly made a lucrative deal, including up-front money and a nice royalty. "What was the big rush?" I asked them afterward. They explained how long it would have taken them internally to get to the production state my model was already in. They even admitted I could have asked for more up-front money. I didn't mind. Knowing from my physician friends that the need for my invention was great, knowing that my invention could be manufactured relatively cheaply and easily, and taking the time and trouble to develop and protect my invention had already paid off.

A final thought about pitching your invention and negotiating your deal: You will have fewer problems if you commit to being scrupulously honest. This may sound naive from a lifelong businessperson, but I'm quite serious. If you go into situations with nothing to hide, business—and life—is much easier. Don't inflate your claims for your invention or for the money you think it can make. If you are thinking that you have to pretend, or outwit, or fiddle with the facts, you constantly have to worry about being found out. It's so much easier to be direct and look people straight in the eye. Give what you hope to get in return—and accept nothing less from the people you work with. If everything about a deal you are being offered sounds great, but the man or woman on the other side of the desk won't look you in the eye and doesn't directly answer your questions, think twice. Be assured that any company president or executive worth his or her salt will be evaluating you the same way.

> As a small businessperson, you have no greater leverage than the truth.
>
> *—Paul Hawken*

chapter five

MARKET WITH A TWIST

You Go to My Head: SKYY Vodka

It may have started with a hungry caveman who'd had a bad day of hunting. Famished, he came upon some rotting, fermenting fruit and made a meal of it. It tasted pretty good and made him feel fantastic. He got such a nice buzz he forgot about the deer that got away and began *waiting* for fruit to spoil before eating it. He'd stumbled upon alcohol.

I've long enjoyed red wine with a meal, or a nice cognac after dinner. I like the way a drink with friends helps me unwind and makes me forget about the metaphorical deer that got away. I find alcohol—in moderation—to be a great stress reducer.

But like many people, I often experienced a downside after drinking. While many suffer a hangover the morning after, I was prone to pounding headaches a few hours after just a drink or two. I mentioned this to my physician friend Martin Sturman. He advised me to stay away from "brown goods" (bourbons, cognac and so on) and to stick with clear spirits like vodka. "They're less irritating," he explained.

So I switched to having a vodka martini or a screwdriver, and I did notice that I got fewer headaches. But one night at a dinner party, I had a cocktail-hour vodka—and a headache before dessert was served. Martin was there, so I cornered him and said, "What's the deal? I'm drinking vodka and still got the headache." His diagnosis: My vodka came from a batch with a high "congener content."

A grown-up version of the kid who once hounded everyone with questions, I pestered my friend for more details. "What are congeners?" I asked. "And what are they doing in my vodka?"

"Congeners are by-products of distillation," he explained. "They give color, flavor and bouquet. They're present in clear spirits to a lesser degree than in the colored ones, but even small amounts can irritate you."

I grabbed a cocktail napkin, asked him to spell the unfamiliar word and stuffed the note in my pocket.

When I got home, I took down my unabridged dictionary and looked up "congeners," but couldn't find it. So the next day I went to a medical library, where I found lots of information about them. I even discovered that there is a Department of Alcohol Studies at Rutgers University that researches just such things.

What I learned is this: During the fermentation of grain, sugar or carbohydrates, a variety of impurities, or congeners, are formed. Ethyl alcohol is what we want, but when alcoholic products are not distilled sufficiently, we end up with amyl, butyl, propyl and isoamyl alcohol, plus acetaldehyde, ethyl formate and methanol. Though you may never have heard the word *congeners,* if you've learned to avoid things like red wine or champagne because they give you a headache, studies have shown that congeners may be to blame.

Much like that concrete wall pulling the fuzz balls off my sweater, this bit of information got me thinking: Could vodka be distilled with fewer congeners? Would such a vodka spare drinkers the headaches and queasiness associated with even a mild "hangover"? Where could I get such a vodka?

I talked to bartenders. They gave me the same advice Martin had: "Stick to clear spirits." I called distillery executives and found that most of them had never heard of congeners. Undeterred, I got hold of a chemist in a distillery lab. I said I wanted a purer vodka, one that would irritate people less. This fellow knew all about congeners, acknowledged that the typical distilling process did produce vodkas with varying levels of impurities and admitted that, theoretically, what I was looking for was possible.

That's all I needed to hear. Like the nineteenth-century inventor of the dishwasher—a woman, who said if no one else was going to invent such a machine, she'd have to do it herself—and as with so many of my other inventions, I made SKYY because I couldn't buy it.

Want, the mistress of invention.

—*Susannah Centlivre, actress and dramatist*

While SKYY is arguably my most successful and well-known invention, I followed the same basic steps in developing it as I had with ideas like the Quad. I observed a problem (headaches after drinking), studied it (congeners were likely the culprit) and realized that solving the problem (getting rid of congeners) would offer consumers a real benefit (less irritation). Once I felt confident my idea was solid, I knew I needed to invent a "model" that *proved* it.

Finding a distiller who was willing to try to meet my standards of purity wasn't easy. In fact, it took a lot of time and persistence. "Who are *you?*" they said when I approached them. They were used to working with the huge spirits companies. I also heard, "We don't need another vodka. There are plenty of vodkas." And over and over I was told, "There's no demand for a cleaner spirit. We've been making vodka this way for years." I finally found one who agreed to take me on—if I paid a premium price "CBD." I'd never even heard that term. It means cash *before* delivery. "Send me a check, and if it clears, we'll work with you," the distiller said. My initial goal had been to invent a unique process that cut the level of congeners to twenty milligrams or less per liter. But once we began to make some real progress, the chemists from the distiller's quality-control department got excited about it and we kept working to refine the process. Ultimately we brought the congener levels to less than ten parts per million. (Today our levels are virtually undetectable.)

When we'd achieved the less-than-ten-milligram quality level, I was convinced I had a great product. I had proved we could create a dependably clean and smooth vodka. But I knew I had no business without the right name. I considered this so important that I sat tight after perfecting my distillation process and didn't proceed any further for almost a year. I just couldn't find a name that clicked. It had to be simple, catchy and convey the difference between my vodka and the others on the market. Without it, there was no point in launching the product.

I have a panoramic view from my San Francisco apartment. One day as I stood at my window looking out at a spectacular, fog-free day, the beauty of the clear, intensely blue, only-in–San Francisco sky hit me. Bingo! That was it: Sky Vodka. It said it all—this vodka was clean and clear as the sky out my window. Harley Procter had had a similar naming epiphany. In 1878, Procter and Gamble's new White Soap was selling well. After a long search for a more distinctive name, Procter rechristened the soap Ivory upon hearing a snippet of the Forty-fifth Psalm ("out of the ivory palaces") during a Sunday morning church service (Panati 1987, p. 218), and greater success followed. Inspiration can strike anytime, anywhere. Amen.

Adding an extra "y" to "sky" gave the name a twist and made it even more distinctive. It worked for Exxon, didn't it? In 1992, nearly five years after my first research into congeners, I was finally ready to take my vodka into the marketplace.

Did I patent my distillation process before producing my "model" or entering the marketplace? No. Though we could have obtained a process patent from the Patent and Trademark Office, to do so would have made our unique process public and after the patent expired, anyone could have copied it. Our proprietary method of filtration is our trade secret. While Coca-Cola's name is trademarked, the script of its logo registered and even its classic bottle shape patent protected, their formula is not patented either. It's too valuable a trade secret. There are lots of wild stories about how Coca-Cola protects their formula. I protect mine by using confidentiality agreements and limiting the number of people who know the nuts and bolts. But the truth is that given the sophisticated technologies that exist today, most substances can be tested and understood quite precisely by anyone who wants to take the trouble to do so. The law offers some protection from this, and allows you to claim trade secret status, if you can show that you consistently make an effort to keep your secret secret. (You need not file for or in any way register your trade secret.)

Getting a license to go into the alcohol business was surprisingly easy—basically a matter of filling out forms. But as usual, I heard "Are you crazy?" more than once. Friends and business associates said, "You can't go up against the big, established guys. They'll squash you like a bug!" I'd cleared the hurdle of producing a vodka that met my stringent quality requirements, and I'd come up with a memorable name, but distributing SKYY—actually getting it into stores, bars and restaurants—was the real challenge. How does an upstart outsider get into the game?

Since I had resources at my disposal, some people might assume I simply threw a ton of money at the problem of distribution, bought a lot of ads and that was it—instant success. It didn't happen this way because while money always helps, money is not what it takes to launch a product. Remember New Coke? Coca-Cola makes more money than many small nations. They were practically paying people to try New Coke, and still it flopped. The truth is, if you have a good product, have read the market correctly and are persistent in your efforts, you can succeed. Money is the result of success, not the cause of it.

I marketed SKYY, when I had money in the bank, the same way I sold the D-Fuzz-It, when I had no money to my name. Instead of getting discouraged when people told me I was making a big mistake and the odds of success were long, I put my faith in my product, got creative and personally delivered SKYY to potential customers to convince them of its worth.

To begin, I hired three employees. My sales, publicity and office people worked out of a very small, no frills office. My salesperson and I hit the streets, visiting restaurateurs and bartenders. I got on my scooter and visited liquor stores. Since regulations set down by the Bureau of Alcohol, Tobacco and Firearms wouldn't allow me to sell on a consignment basis, I went in and made a personal pledge: "If you buy three cases at the wholesale price, I'll come back in a week and personally buy back at full retail price any bottle you haven't sold." In other words, I wasn't taking "returns." I was offering to buy the product just like any other customer. I believed in SKYY that much, and that kind of belief is contagious. I never wound up having to buy a bottle of SKYY at retail.

We spread the word about congeners and our nearly congener-free vodka. Bartenders were very receptive. They like having something exciting to share with their customers. The style- and health-conscious San Francisco gay community took to SKYY immediately as well. You can't buy loyalty—you have to earn it. People may want to try whatever seems trendy and new, but they won't come back for more unless you make good on your promises. We did.

In this personal, one-to-one way, we forged relationships and made converts. A good name and a nice sales pitch can't hide a weak product. People can see through hype. If you have an honest, innovative item at a fair price—since SKYY is made in America, we can sell it for significantly less than the premium imports—you can sell it. We knew we had something solid to offer, and just like those first D-Fuzz-Its, our initial bottles of SKYY proved our point. We got orders.

Still, a little free publicity couldn't hurt. The story of congeners and how they relate to that hangover feeling had been news to me. Where do you go with news? Newspapers. News is their *business*. We wrote a press release that told the SKYY/congeners story and sent it to the local papers. In June of 1992, the *San Francisco Examiner* ran with it. A staff writer interviewed me and wrote an attention-grabbing story under the headline, "Inventor shoots for pie in the SKYY." They even sent out a photographer and included a photo of me with a bottle of SKYY in the story. Readers' curiosity was piqued and they began asking for SKYY by name.

The only distributor we could get to work with us at the outset was a small imported beer distributor. We were not his top priority but as the publicity increased the demand for SKYY grew, and there was a domino effect. The distributor got more orders.

USA Today picked up the story ("Vodka purified to cut hangovers"), followed by newspapers and magazines

Sunday
JUNE 14, 1992

Inventor shoots for pie in the SKYY

Kanbar's vodka allegedly causes less pain

By Robert Slager
OF THE EXAMINER STAFF

NOT TOO many good ideas come after drinking a couple of screwdrivers. Oh, you may think so at the time, but a decent headache has a way of replacing such thoughts. Maurice Kanbar felt that invisible vice around his head about two years ago at a cocktail party, but for him, a thought came popping out.

Kanbar, an inventor by trade, considered the possibility of purifying vodka to the point where it would no longer cause a headache. Kanbar believed there were other "sensitive drinkers," that could be enticed by a vodka that was free of congeners, the toxic impurities that form during the fermentation process.

Using a distillation process in which the vodka is boiled at four different temperatures, Kanbar has introduced a new product to the Bay Area market: SKYY Vodka.

"I just couldn't handle the side effects of certain alcohols," Kanbar said. "I love the feeling of drinking in moderation, though. It's probably the best natural sedative given to man, but I just couldn't drink brandy or bourbon because of the headaches I'd get. And then when a white spirit like vodka gave me trouble, I became curious with the possibility of purifying it."

Kanbar, born in New York and a graduate of the Philadelphia College of Science and Technology, said his new process doesn't alter the alcohol content of the drink, just eliminate most of the congeners. The process isn't intended specifically for heavy drinkers, he said. Some people can get headaches from just one drink. And he isn't promising a hangover-free morning if you drink mass quantities (you will still be dehydrated), but he said the headache will most likely be gone.

"Not everyone gets headaches from drinking, of course," he said. "Some people even think that congeners take some of the edge off the alcohol and give a drink character. This is intended for the 3 percent or so that are affected by the congeners."

Kanbar said he is always looking for the big breakthrough. He already has 30 patents to his credit including the "De-Fuzz-It," a little comb-like device that removes fuzz balls from sweaters. It may not sound like much, but Kanbar said sells about 300,000 a year to discount stores across the country.

He also has patented several surgical instruments including one for cataract cryogenic removal, where the lens of the eye is frozen and removed intact. He sold that to Alcon Laboratories (a division of Nestle) in Austin, Texas. He also sold a type of surgical retractor used for varicose veins to Zimmer, a division of Bristol Myers.

"I look around all the time and just have a feeling that everything can somehow be improved in design," he said. "I mean, think of it. When we were kids we used to underline books with a pencil. Then some genius thought of highlighter pens [illegible] thought up Post-its. What brilliant ideas. Post-its did about $230 million in business last year. You don't have to be technologically brilliant to do this stuff. You just have to be alert."

Kanbar, single and living in The City, hopes the next great discovery will be his San Francisco-based SKYY Vodka. Distilled in Illinois and shipped to a bottling plant in San Jose, SKYY Vodka has found its way to about 80 locations in the Bay area including Pat O'Shea's on Geary, The Blue Light on Union, The Lone Palm on Guerrero and the Lark Creek Inn in Larkspur as well as several retail outlets. It's being distributed by Chrissa Imports in Brisbane.

Jeff Jordan, owner and general manager for both the Blue Light and the Fillmore Bar and Grill said SKYY Vodka is slowly becoming a seller.

"The taste is real smooth," he said. "The response has been very positive. It's still a call drink (needs to be requested specifically), but we had a SKYY Vodka night a few weeks ago where we used it for all Vodka drinks. We've been getting more and more people coming in and asking for it specifically. It's pretty competitively priced."

Word-of-mouth will have to be the primary marketing tool, Kanbar said, because the advertising budget is still very small. With four employees, SKYY Spirits Inc. first began marketing in San Francisco because "it was a large enough area to gain a following but still small enough where we could track the progress," Kanbar said. He also noted The City's reputation for quality living and high regard for good health.

Kanbar doesn't expect to turn a profit for at least two years. He can't patent the process, he said, because it is a concept and not an actual invention, but he sees big things ahead.

The plan for now is to move to Los Angeles in six months if all goes well. If the vodka takes off in California, Kanbar hopes to take it to New York.

That would bring him full circle. Kanbar, who is also on the board of directors for the San Francisco International Film Festival, grew up in New York and [illegible] have let a big money venture slip through his fingers.

"In 1972 I owned a building in New York and decided that what New York needed was a fourplex cinema. I coined the concept, but didn't expand it and didn't earn any money with it. I was just happy to have these theaters because I loved film so much. It's tough to look back on that now."

If his SKYY Vodka can make its way back to New York, it just might eliminate the bad taste in his mouth.

EXAMINER/MARK COSTANTINI

Maurice Kanbar *believed there were "sensitive drinkers" who could be enticed by a vodka that was free of headache-causing congeners.*

from the *Wall Street Journal* and *Newsweek* to *Details. Business Week*'s story was headlined "Hangover-proof 80 proof?" I was invited to appear on local and national TV news programs on networks including CNBC and Fox. Before we knew it, David Letterman was grabbing a bottle of SKYY that a guest chef was using in a cooking segment and "guzzling" it prominently on his *Late Night* television show. More orders.

In less than ten years SKYY has achieved annual sales of more than $50 million. We've gone from talking shopkeepers into giving us a square foot of floor space in their store to being courted by distributors anxious to do business with us. SKYY is now sold in every state in America and in eighty countries. We've raised the bar for the entire spirits industry, with competitors eyeballing our quality standards and imitating our marketing techniques. I guess you could say we shook things up.

SKYY's success is rooted in the quality of the product itself, but naming, packaging and marketing also play a crucial role, as they do with any product. Once you have something worthwhile to sell, you need to convey that value to customers, preferably with a twist as distinctive and memorable as your product itself. Let's look at how we did that with SKYY.

First Impressions: Name, Package, Price

What's in a name? Everything. Was *Dirty Dancing* a successful movie because it was a great film? No. The title brought people into the theater. I doubt they'd have come out for the same film if it were called *Weekend in the Catskills*.

Although I've never done it, I've often said that if I had a great name for a product, I'd build a business around the name. That's how important names are.

In 1965 I got a patent for a flat film dental floss that I'd invented. It slid between teeth easily and comfortably. I tried to sell it to Johnson & Johnson and Colgate but found no takers. I thought about manufacturing and selling it myself, but I couldn't come up with a snappy name—I kept calling it something clunky like Film Floss. I recently saw a new flat film floss on the market called Glide. That's a terrific name. If I'd thought of it, or maybe even just Slide, I would have manufactured the stuff myself. Alas, my patent is long expired.

Unlike patents, names don't expire. That's one reason why they are so valuable. While the value of a patent tends to depreciate, since competitors can study your patent and begin to build on it, names appreciate. No one else can use them, and over time they attract recognition and loyalty.

Sometime in the 1970s, long before you could find Evian and San Pellegrino at your corner store, I was eating at a restaurant in a small Georgia town. A man and his date were sitting at the table next to me and the man asked for a Perrier—not a "bottled water" or a "club soda," but a Perrier. I'd bet that the fellow was trying to impress his date with his discernment and sophistication. A French name that's easy to pronounce achieves a kind of cachet that has tremendous value for the manufacturer. People feel good about themselves when they buy the product, and other people notice and follow suit because they want that touch of class as well. It's brilliant marketing.

Names like that become part of the language. Do you make a copy or a Xerox, use tissues or Kleenex, apply petroleum jelly or Vaseline? Those names are like gold.

Legally your product name is trademarked the first time it is used in public. You can use ™ after your name from the first date of use. But you can use the ® symbol only after you've registered your name with the U.S. Patent and Trademark Office. This is simply a matter of filling out paperwork. The office then reviews your trademark and double checks that it is unique. You can file either a "use" application, if your trademark has already been commercially used, or an "intent-to-use" application. Trademarks may be registered only when they are used in interstate commerce; you need to show that you've sold the product to someone in another state. And commonly used words can't be trademarked; that's one of the reasons for SKYY's spelling. Trademark registration must be renewed every ten years, and a mark may be considered abandoned after two years of nonuse.

> A good name is rather to be chosen than great riches.
>
> —*Proverbs 22:1*

To make sure that the trademark you've chosen isn't already in use, you should do a search of both registered and unregistered marks. There are trademark search firms, directories are available in most libraries, and there are computer search programs. Unregistered marks can be researched in trade directories as well. (See appendix.)

During the months when I was trying to name my new vodka, I toyed around with things like Prince Nikolai or Czar Alexander—obviously hung up on the Russian thing. But how many people would go to a bar and say, "Give me a Czar Alexander and tonic"? It's too complicated and sounds fussy.

Your product name has to sound right in just these kinds of real-life scenarios. You have to imagine your name being used in the situations your customers will find themselves in. I could hear that fellow in Georgia ordering a SKYY martini, SKYY Mary, SKYY-driver, or SKYY-hound, just the way I could hear a woman saying, "Let's go to the Quad." The right name clicks. (Unfortunately, I neglected to register "Quad." A rash of four-plexes sprang up after my success with the concept, and if I had registered the name, I could conceivably have collected licensing fees from my imitators. Learn from my mistake.)

Consider what image you want to convey. Clinique and Prescriptives are cosmetic companies whose names signal products with scientific benefits. But Urban Decay is also, believe it or not, a line of cosmetics. They feature nail polish colors like Uzi, just right for their target market: the Irony Generation.

I actually had the name "Vermeer" before I had perfected the chocolate cream liqueur that would bear the name. I knew I wanted a rich, even sensuous drink. If you've ever seen a painting by the seventeenth-century painter Johannes Vermeer, you'll see the connection. Vermeer's work is warm, golden and suffused with light—it's every bit as sensuous as our drink. Plus Vermeer was Dutch, like the premium chocolate we use—and I've read that he was just as obsessed with details as I am. When I'm in New York, I regularly visit the Vermeers in the Frick Collection. We even recreated a detail from the Vermeer painting *Girl With a Pearl Earring* on our label. (Sometimes you just get lucky. I decided I wanted to call my new chocolate cream liqueur Vermeer three years ago, and was warned that a lot of people might not have heard of the painter. Suddenly in the last year, there have been four very successful books concerning Vermeer. The best-selling novel *Girl With a Pearl Earring* even features the same painting on the book jacket as we're representing on our label. The serendipity grew when an opera, *Writing to Vermeer,* by Peter Greenaway, played at Lincoln Center the week before we launched.)

Ordering Perrier at a bar or restaurant didn't become a classy and health-conscious thing to do in the seventies and eighties because of the bubbles. Its spring source, which is thought to be 130 million years old, was originally called Les Bouillens, or boiling waters. Napoleon III granted the rights to commercialize the spring in 1863, and the bottler had the good sense to call it Perrier, the name of the physician who headed the Société des Eaux Minerales, Boissons et Produits Hygeniques de Vergèze. When the venerable product exploded on the U.S. market in the mid-1970s, it capitalized on both its European sophistication and its natural, calorie-free health benefits. The company emphasized its "hygeniques" by sponsoring the New York Marathon and Los Angeles Olympic Games, and by creating its "Earth's First Soft Drink" tagline. Perrier's sleek, elegant and instantly recognizable green bottle added to its allure.

The first safe commercial hair dye was marketed in 1909 by the French Harmless Hair Dye Company. At first, the company moniker was also the product name, and while this conveyed important information about the product's quality, it obviously lacked something. A year

Tips for Brainstorming a Name

- *Make up a word (Oreo, Kodak).*
- *Consider positive image words (Eternity perfume, Renew recycled trash bags).*
- *Use a unique spelling of a common word (SKYY), or a combination of words (Slim-Fast, Pull-Ups, Band-Aid).*
- *Speak a foreign language, or pretend that you do (Tavolo.com, Frappuccino).*
- *Use scenarios—what will someone say, notice, remember?*
- *Ask your friends—I overrode a friend's Movies 4 suggestion but another friend's idea spared the world the Balls Off sweater comb.*

later the name was changed to L'Oréal, and suddenly you knew your hair wasn't going to be just safely dyed but also glamorous (Panati 1987, p. 233).

Vermeer, Perrier and L'Oréal add a touch of class, but there are no hard-and-fast rules for names. You can be more literal to convey a product's key benefits, such as Huggies Pull-Ups, Slim-Fast, Good Grips, and Baby Jogger. These names are snappier than "French Harmless Hair Dye," but their focus is still usefulness, concentrating on product features rather than glamour or style. With some products and in some markets, this is more appropriate.

I wasn't just exacting about naming SKYY. It also took me months to come up with a better way to say "Made in America." I wanted to let people know that my vodka wasn't Russian or Swedish, but the way I said this was important to me. The phrase had to flow and have the right cadence. I finally came up with "Distilled in America from American Grain." Doesn't that sound better than "Made in the USA"?

These aren't minor details. They are an integral part of what makes a brand distinctive. SKYY is not just guaranteed to contain no more than .01 grams per liter of total congeners, it has also got a simple, memorable name and it's American. You have to give people an incentive to try your product, and then make it very easy and appealing for them to stick with it. Give them reasons to explain their choice, make them proud to order it. When people ask for SKYY, they show that they care about product quality and that they understand the difference between premium vodkas.

➤ *Perfecting the Package*

If the name of the vodka is SKYY, you can hardly put it in a green or yellow bottle. As night follows day, the bottle should be blue.

But at the outset, I was cautious. Using a blue bottle wasn't going to be easy or inexpensive, so I wanted to be sure of several things: Did I have a truly fine product? Yes. Was there a market for the product?

To find this out, I distributed five thousand cases of SKYY in a water-clear glass bottle in the Bay Area, and spread SKYY's low congener/antiheadache story. I had to prove that there was an audience, that people were listening. By pounding the pavement, conducting taste tests, and telling our story, I learned that drinkers did understand the value of SKYY enough to buy it in an unremarkable bottle. And our initial bottles were truly plain. Sometimes after Dave Stoop, SKYY's first salesperson, and I did a taste test with the bottle concealed in a brown paper bag, the potential customer would say, "You're right, it's great. Now let me see the bottle." Some of them grimaced when we revealed our generic, almost medicinal bottle. But I was determined to test the value and salability of SKYY on its own merits.

Every Friday afternoon, I met with my employees in my office to recap the week's events. Invariably the plain bottle would come up. "*Please* get us a better bottle!" one would say. "First sell five thousand cases," I insisted. Finally, when I realized our initial customers were loyal, I decided to pull out all the stops and do whatever it took to get a blue bottle. Our limited, local initial sales and taste tests were essentially "end-user" trials. Our feedback told us that our customers liked everything except our bottle.

Still, because I was about to break with industry standards, I heard "Are you crazy?" all over again. Some people said, "You can't see the level of vodka in a colored bottle" or "No one uses a colored bottle for a clear spirit." Exactly. No one else was doing it. I knew the blue bottle would make SKYY even more distinctive, and I was willing to bet that any minuses would be outweighed by the plus of having SKYY truly stand out behind the bar or on the store shelf. I clung to my idea like a pit bull.

What's more, when I hear a businessperson say, "That's the way it's always been done," I bristle. If you've been doing something the same way for twenty-five years, I wonder why you aren't innovating—and I know that you're leaving yourself vulnerable to a competitor brave enough to try something new. Most of the time, when people say, "You can't do that," they mean, "It's never been done." That's not a good enough reason.

> Business has only two basic functions—marketing and innovation.
>
> —*Peter F. Drucker*

Where would Apple be if they didn't "think different"? If their iMac design team had said, "Computers are beige. We can't make colored computers because no one ever has," Apple probably wouldn't be enjoying their current revival. Of course, the iMac also had to be easy to use and competitively priced, but it's the translucent plastic colors that embody their cutting edge style with the public.

Jonathan Ive, the iMac industrial design team leader, has said that his team approached their project with no preconceived notions of what a computer should look like. Before he was an Apple employee, he had been devoted to the Macintosh because of its "more emotive, less tangible product attributes." He brought that history to the iMac project, creating colors not just to differentiate the product in the marketplace, but also to "create products that people would love." Setting trends isn't incidental or accidental. At Apple, it's their MO. As Ives states, "In a company that was born to innovate, the risk is in not innovating" (Hirasuna 1998).

Despite my determination, obtaining a blue bottle for SKYY was difficult. I could not find an American glass company willing to manufacture one for me. I learned early on that you should never refuse to do what a customer asks, even if it seems unreasonable. Rather than say, "No, that's too expensive" or "No, that's too much trouble," you should listen to the customer and then say, "This is what it'll take." Charge them for your time and trouble, of course, but do what they ask. You never know, if you go the extra mile for them, they might turn out to be your best customer.

I found a company in Hannover, Germany, that was making blue bottles for another customer. I had to pay them much more than I was paying for clear glass in the United States, but I was willing to do so and they were willing to work with me. We still get our bottles overseas—millions of them—and the stateside companies lost out on a big customer.

Just as Coca-Cola is recognized by its bright red label, we are identified with our distinctive cobalt blue. Not only does the bottle play off the name (sky blue), it also emphasizes our lack of impurities (clear blue), and it stands out on the shelf. We introduced the blue bottle in 1993 and a year later it was featured on the cover of *Packaging World* magazine with an inside story headlined, "SKYY Gets It Right." Two years after launching, we were enjoying 15 to 20 percent jumps in sales per month, due in no small part to the bottle improvement.

Imitation may be the sincerest form of flattery, but when new vodka brands were launched in blue bottles I was not amused. "Trade dress" is a distinctive but nonfunctional feature that distinguishes one product from others. As time goes by, unique and distinctive trade dress, like a trademark, becomes more valuable as a means of identification for the consumer. Also like a trademark, trade dress can be registered with the PTO. But it doesn't have to be in order to receive protection. If it's distinctive and consistent, and if an imitation could cause consumer confusion, unfair competition law—an admittedly gray and changeable area of the law—gives you offensive rights. If you bring suit, a judge could tell the other business to stop or modify their behavior and even award the injured business monetary compensation for lost profits.

For example, if I decided to go into the photographic film business, I couldn't package my film in a yellow box. That would invade Kodak's trade dress. Kodak owns that look. They could easily make the case that even if my film was called Glotz, if I packaged it in a copycat yellow box,

I was attempting to confuse consumers into buying Glotz when they wanted Kodak.

One of our blue-bottle copycats called themselves Ultraa—they even copied the double letters in the name! But their bottle's shape was quite different than ours; it was curvy and voluptuous, so I didn't take any action. Another cobalt blue imitator used a bottle that at first glance could be mistaken for ours. I felt that was a trade dress infringement and was willing to go to court over it. Chances are good that a judge would have ruled in our favor, but we settled when the competitor agreed to sell out their current stock and then stop copying us. There's a time to be flattered, and trust that your customers will remain loyal, and there's a time to take action. Consumers had come to expect to find SKYY Vodka in cobalt blue bottles, and we had a right to protect that.

I could have hired a designer to create our new bottle, but I didn't. I have nothing against designers, not at all. But I'm picky and have a strong, gut-level sense of what does and doesn't work. At the beginning, with the D-Fuzz-It for instance, I couldn't afford designers. I did my homework, went with my instincts, and the result was successful. That trained me to trust my gut and reinforced my do-it-yourself tendencies. Most recently, for the Vermeer packaging, we did work with a designer. I had the name and the product image, and knew I wanted to recreate a portion of a Vermeer painting on the label, but I also wanted to stretch out and experiment a little. I felt I could afford to see what someone else could come up with and then direct the changes. The collaboration worked out very well.

But with SKYY, I looked at various bottle shapes and chose an extremely simple, slim bottle with sloped shoulders. SKYY is a direct, honest product. The bottle needed to reflect that simple elegance. If something is fussy and ornate, you are more apt to tire of it. Plus, an intricate bottle would have contradicted our clear and clean product. To position your product in the marketplace, and garner interest and excitement, the product, name, package and promotion should fit together seamlessly.

Speaking of bottles, until 1915, Coca-Cola came in a bottle that was very similar to those of other soft drinks. To give Coca-Cola drinkers a way to spot the product before their first sip, the company replaced their generic straight-sided bottle with their contoured one that could be recognized "even if it was felt in the dark" (www.coca-cola.com). In 1977, in a move rare for the Patent and Trademark Office, Coke's unique bottle was granted registration as a trademark.

I did have a designer present me with some label mock-ups for SKYY, but nothing she came up with felt right. So I went the D-Fuzz-It route; that is, I chose the typeface, type sizes, and colors myself. But I did have one hell of a time getting the gold I wanted for our label.

In Jim Jarmusch's recent movie *Ghost Dog,* a character talks about a tenet of samurai philosophy: you can put aside the really big problems but the small problems require a lot of attention. Next time someone tells me I'm too finicky, I'll tell them I'm a bit of a samurai because I have always found that the "little" things add up and matter.

I kept telling the printer we were working with on the SKYY label that I wasn't happy with the quality of the gold lettering. He thought I was nuts. (Join the crowd.) The problem was that if you looked at the label from a certain angle, the gold took on a distinctly green tinge. I didn't want that. The printer said there was nothing he could do about it. "It's fine. No one is going to look at it from the side." But *I* looked at it from all sides, and it bothered me.

One day I noticed a gold can of caffeine-free Coca-Cola. I had learned that that look is achieved with transparent yellow ink. You can take paper and vaporize, or vacuum deposit, a very thin coat of aluminum on it and the paper ends up looking metallic. If you print over that with transparent yellow, it looks gold. If the yellow has some red in it, it looks like copper. The Duracell copper battery is in a steel case that is printed with a transparent yellow-red so that it looks like copper.

So I'm holding this can of Coke and thinking, "Now there's a gold; there's no green in *that*." With this validation that what I wanted could be done, I made some calls and found the manufacturer of the ink Coke used. Inks, Inc., turned out to be just across the bay from me, in Oakland, California. Confronted with the evidence, my printer said, "Okay, okay. You're right. We can make that gold for you."

Never leave well-enough alone.

—Raymond Loewy, industrial designer

But that wasn't the end of my label woes. Our first labels got scuffed by the cardboard separators in their boxes and in normal handling. I was told this was to be expected, that it was no big deal, and that it would only happen to one out of every thirty or forty bottles. What's the big deal if two or three out of a hundred bottles are imperfect? The trouble is, if a person picks up a bottle, sees the scuff, and puts the bottle back on the shelf, you'll eventually wind up with six scuffed bottles on the shelf. Remember to think even the smallest things through to their logical conclusion—do the scenario. I insisted on scuff-resistant labels.

I first tried having the labels coated with a special transparent ink. But in order to become tough, the ink has to be exposed to ultraviolet light; the ink literally polymerizes and forms a thin plastic coating. The folks who did my coating did it so fast that it didn't really polymerize and didn't have the abrasion resistance that I wanted. After some more back and forth, I decided to laminate a thin film onto the label after it was printed. It's a little more expensive, but it works to protect the label.

All of this label perfecting probably took nine months. It's easy to be tempted to let some of these details go in favor of getting your product out there. But if you let them go, they have a tendency to compound until you are left with a second-rate product or package. You want to end up with the quality and value you initially imagined—and you can if you're a samurai about the details.

If you can't draw and "don't know a thing about design," you can simply borrow from the best. I don't mean infringing on another product's trade dress! I mean what I did when I went to drugstores and studied cosmetics packaging in order to get ideas for merchandising the D-Fuzz-It. As consumers, we all know more about design than we think we do.

Pay attention to what attracts you in ads, in the grocery store, at the drugstore, everywhere. What colors catch your eye? What's in style but is classic enough to outlast fashion? What kind of type grabs you and gets its message across? Whenever we prepare to launch a new product, the people who work in my product development lab go out and do exactly this kind of research.

When you see something clever, make a mental note. As we've already discussed, a great deal of inventing is building on what's come before, not reinventing the wheel. Study what works and, as TV chef Emeril Lagasse says, "Kick it up a notch."

> *Consider placing a bulletin board notice up at a local art school or college. You might be able to obtain high-quality design help at reasonable rates from a student.*

➤ *Making Sure the Price Is Right*

Like your product's name and package, its price is also a key factor in making a positive first impression.

You might think that a product's price is based solely on what it costs the manufacturer to make, distribute and market. But other factors come into play. While I believe in focusing less on what the market will bear and more on making a fair profit and delivering honest value, there are other strategies.

We've talked about the fairly standard 4-to-1 markup, and how a company's low or high overhead can affect that. But industry standards vary. Cosmetics often retail at ten times their cost due in large part to their massive advertising budgets. And products are sometimes priced high just to give the illusion of greater quality. It's not my practice, but it happens all the time.

The two most common strategies are to price low, in order to capture market share, or price high, to convey exceptional quality. If you price low, you hope your reduced per unit profits will be offset by increased sales—and that your buyers will remain loyal if you subsequently increase your price. If you price high, your initial sales may be smaller, but they may be made to more influential buyers—and if you subsequently drop your price, you'll be perceived as offering a great deal.

People want economy, and they'll pay any price to get it.

—Lee Iacocca

Think about how you make buying decisions. If you're like me you sometimes go for particular brands no matter what their price, and at other times buy whatever brand is on sale or priced the lowest. You weigh price against value. If you are convinced that a particular laundry soap makes your whites whiter, your brights brighter and smells like springtime, you'll probably be willing to buy it over and over again even if it costs a bit more. But if you think any old laundry soap will do, you'll buy the cheapest one. When you're selling a product, if you can't convince consumers that the value you offer is worth a price that allows you to make a profit, you're out of business.

Your product and market research should have given you a clear idea of pricing standards in your field. You might also have polled people on how much they would be willing to pay for the product they were testing.

As noted above, while SKYY is a premium vodka, we can sell it for less than the other premium brands because we manufacture domestically. We see and will position Vermeer as an affordable indulgence—an "everyday luxury" like champagne or a good wine, with a price comparable to that of Baileys Irish Cream.

As usual, solid research is essential in determining a price. Study and know your market and your consumer. Understand your product's essential value and work to manufacture and deliver that value in a way that affords you an honest profit.

What's the Buzz? Spread-the-Word Distribution and Marketing

We've seen that distributing SKYY posed a real challenge. Unless you license your invention, or sell and distribute it yourself, as I did with the D-Fuzz-It, obtaining good distribution will be one of your challenges as well.

As I said, our initial SKYY distributor was a small local imported-beer distributor who agreed to add a case or two of SKYY onto his truck. We simply couldn't get anyone else to take us on. The San Francisco Bay Area was our initial test market and between that smallish market and our lack of a track record, the big distributors we approached turned us away with, "Why should we have our sales force promote you when we have Absolut?"

Our distributor didn't have a sales force per se, so Dave Stoop and I made sales and faxed the orders in to the guy with the truck. As we've seen, my employees and I also worked nonstop to build our presence. As people heard our name, saw us at events, and read about SKYY and congeners in newspapers, our distributor began to get calls and orders.

Buoyed by our good results, we moved to statewide distribution and this time hooked up with a wine distributor. By now we could photocopy articles about SKYY that had appeared in the press and give our sale reps this additional ammunition for sales calls.

You can't depend on distributors to do it all for you. Please read that sentence again; it's that important. Distributors sell, and their sales forces get behind, products that sell. YOU make that happen. For SKYY, that meant cultivating word of mouth, being visible at events, and earning publicity with the SKYY story. (We didn't begin advertising until we were already a success.)

Make your customers de facto sales reps, or as Guy Kawasaki might say, evangelists. Bartenders played that role for us. When we convinced them of SKYY's value, they spread the word to their customers, who then bought SKYY at their local liquor stores and called for it by name at other bars and restaurants.

No matter how established you are, distributors can't afford to handle stuff that doesn't sell. They will give Vermeer a shot because of our SKYY track record, but they won't stick with Vermeer unless we do our part by promoting the product.

When we made the switch to cobalt blue, our sales got quite a boost. When we went to the annual Wine and Spirits Wholesalers of America trade show in New Orleans that year, distributors approached *us*, wanting to work with us on national distribution because they had heard how well we were doing.

When it came time to distribute SKYY in Europe, we faced distribution issues all over again. I didn't want to set up remote branch offices and hire and manage our own people abroad. The solution lay in the forming of strategic alliances. For instance, Campari International sells well throughout Europe and Asia, and has an in-place infrastructure and distribution network. We made a deal so that Campari represents SKYY overseas and we are now their exclusive U.S. importer. In a short time, Campari found themselves selling twice as much SKYY as they'd forecasted in Italy—not traditionally a vodka-drinking country. *Salute!*

Snapple's David and Goliath story bears some resemblance to SKYY's and is worth mentioning in this context. Founded by three entrepreneurs from Brooklyn, Snapple developed a solid base in New York City health food stores and then gradually expanded their product line and distribution into supermarkets and delicatessens.

In order to implement their idea for a preservative-free, real-brewed ice tea, they innovated a "hot fill" bottling process similar to home canning. This highly successful tea attracted large distributors to Snapple.

The nonalcoholic carbonated and noncarbonated drink business is dominated by a few giants. In 1994, Coke and Pepsi held 72 percent of the entire market. Superpower Pepsi partnered with Lipton Tea to compete with Snapple and create a bottled tea, but, as is often the case with gigantic companies, it took them several years to do so. (Unencumbered by plodding, bureaucratic research, development and decision-making processes, small entrepreneurial outfits can often out-pace larger, established organizations.) But when Pepsi-Lipton did fight back, they fought hard, questioning Snapple's brewing process and using the full weight of their distribution network to launch their Lipton Original everywhere. Snapple beefed up their advertising but also benefited greatly by having been first to market with their tea. Plus, Snapple's niche—natural "New Age" beverages—became a growing market trend. Finally, and also like SKYY, Snapple brought top industry professionals into their operation, crafted exciting consumer advertising, and launched its line in Europe, all while staying true to their initial quality and image (Thomas 1995, pp. 290–97). But let's get back to the basics.

There are an increasing number of ways to get your product to its consumer, and as usual, your imagination, creativity, and enthusiasm are key to making any of them work. Routes include:

- Stores
- Catalogs
- The Internet
- Infomercials
- Direct Mail
- Use As a Promo Item

Let's look at a few of these options.

You may be surprised to learn that you can pitch your product to huge retail outlets like Wal-Mart. Find out when and how they see new products and make your case. (See appendix for contact information.) If I were bringing my SKYY Timer in for consideration, I'd do a little something extra like bring in a dozen printed with "Wal-Mart Time" instead of "SKYY Timer." I'd hand one out to everyone I could, knowing that using the device is the best way to discover its worth.

The ROLLOcane is currently selling through a small mail-order catalog that handles all kinds of similar gadgets. Maybe you've got something The Sharper Image or Hammacher Schlemmer would be interested in. Both of these retail and catalog outfits regularly look at new invention/products and Hammacher Schlemmer even runs a Search for Invention contest that rewards inventors of patented products that have not yet reached the marketplace. (See appendix for details.)

As I previously mentioned, a savvy business associate recently suggested I do an infomercial to sell the ROLLOcane. It's not a bad idea. Many products have been known to move on into established retail markets after succeeding in the world of infomercials. Some infomercial production companies will even look at a product in the design stage, and if they like it, handle manufacturing.

Another avenue I've been considering for my SKYY Timer is offering it to another company for use as a promotional item for them. David Pressman cites the famous example of eight-year-old Abbey Mae Fleck who invented a plastic device for hanging bacon in a microwave so that the grease dripped away while it cooked. She couldn't get any manufacturers of microwave accessories interested and so she had the very bright idea of approaching a bacon company. They put a discount coupon for her product on their packages and it was a huge success (Pressman 1997, pp. 11/13). Like SKYY, almost every kind of company uses promotional items that can be emblazoned with their name and either increase their sales or simply keep their name in the public's eye.

No matter what distribution channels you pursue, remember that distributing is a business and, like all businesses, needs to be profitable in order to survive. Distributors don't exist solely to help *your* company survive. So show them what's in it for them. Persuade them that your product will generate inquiries and sales. Convince them that you will promote the product and then do it.

Publicity, Promotion and Advertising

I'm told that it is rare for an inventor, who often has a science or engineering background as I do, to also be interested in marketing, which I am. I actually find the two areas complementary. Successful inventing is every bit as creative as good marketing, and it pays to approach marketing logically.

Many inventors have also been hands-on marketers. Robert Augustus Chesebrough, the inventor of Vaseline, was one. While I scootered around San Francisco, visiting liquor stores with SKYY, he traveled around upper New York State in a horse and buggy, giving jars of Vaseline to anyone who would take it. His version of my daily SKYY martini was a spoonful of Vaseline each and every day. He claimed he lived to age ninety-six because of his daily dose, and I hope my ritual does the same for me!

While it's certainly possible to stay in your lab or work room and hire people to package and sell your inventions, I find marketing exciting and fun. No one understands my inventions better than I do. But if marketing is not of interest to you, sales and marketing associates can be people you hire to be part of your company, outside experts your company contracts with or a division of the manufacturing firm you license. Your company can also concentrate on sales and marketing while subcontracting the manufacturing. The choice you make depends on your interests and skills.

Dave Stoop was a kid studying at San Francisco State and working as a caterer when he heard about SKYY. He thought it was a great idea and approached me about part-time work. I talked him into working full-time as my first salesperson, and he soon became known around San Francisco as Mr. SKYY.

In our office, four people had to fight for the use of the three phone lines but Dave was rarely there; San Francisco was his office. He visited liquor stores by day and restaurants by night, where he'd drink one cappuccino after another while pouring taste tests for bartenders and restaurateurs.

We did our taste tests against the top-selling premium brands—Absolut, Stolichnya and Finlandia. We nearly always won. This one-to-one selling was part of our low-budget strategy for building a market presence.

Because our bottle was so attractive and so different from any other, we were able to do striking window displays in stores. The way light hits the bottle is just spectacular, so merchants were happy to have our displays up for months on end.

When I was twelve and growing up in Brooklyn, my best friend Harvey and I started a business, HarMor Photography. Basically we took pictures on spec. If the neighbors said they didn't want to have their picture taken, we said, "Smile!" and took the picture anyway. When we showed them a print, they could rarely resist buying a copy. When a restaurant, store or bar resisted selling SKYY, we persevered, giving them a taste test, creating an attractive display or offering them a risk-free investment. Nine times out of ten, when they saw the results, we were in business.

Aside from SKYY itself, we had a brochure I'd written, "Whyy SKYY?" that laid out the science and innovation behind the brand, plus T-shirts and key chains with our logo. We left those promo items everywhere we went—phone booths, cabs, even restrooms. Through the years, SKYY became known for unusual, useful giveaways. We had SKYY wristwatches made in the style of some of the most popular brands. I once asked the fellow sitting next to me on a New York to London flight for the time. He looked at his SKYY watch. I asked who gave it to him and it turned out to have been my nephew who was marketing SKYY in the Hamptons. (We'd targeted that upscale Long Island enclave when we brought SKYY out East.)

We try to use promo pieces that have some value to the customer—nothing intricate or expensive, just fun, handy, branded items that people will keep and use. Vision is an interest of mine so I put a chart that tests for color blindness on the back of our brochure. If you can read the letters S-K-Y-Y, you have good color vision. People love to test themselves with such things and it's a fun thing to show a friend—especially one who makes weird color choices!

We make SKYY martini atomizers (vermouth spray) and did a co-promotion with a gourmet olive company to create SKYY boxes filled with a metal shaker (with a SKYY logo, of course), two glasses, an atomizer and a jar of olives.

We also have something called SKYY Eyes, strips of plastic with tiny pinholes that, when held up to the eyes, bring blurred images into focus (they also function as swizzle sticks). I was sitting at home one day, listening to a local radio station. The talk-show host mentioned that he'd forgotten his glasses and was having a hard time reading his materials. I got on my scooter, rode to the station across town, and left a pair of SKYY Eyes and my business card with the receptionist. The talk-show host got a big kick out of the glasses—and gave SKYY a nice plug.

Our distribution widened and our sales skyrocketed when newspapers, magazines and television began telling our story. No amount of advertising could have the impact of a story in *USA Today* or a five-minute interview on CNBC. Lots of these media outlets were skeptical about our claims. They grilled me and went out and conducted their own taste and morning-after tests. We passed with flying colors, and I had the science to back me up. We can't promise "purity" or guarantee no "hangovers," but people who tried SKYY were pleased with their results, and these news stories got more people to check us out.

Our experience with the press is another instance of how valuable having a legitimate story to tell can be. Not every new product is newsworthy and the media won't gratuitously plug products. But if your product or invention is "fit to print," if it can make a provocative claim like SKYY or if it's the first of its kind like the Quad, bring it to the media's attention with a press release.

As Bob Coleman and Deborah Neville point out, in 1991 Ben & Jerry's made it onto the *Wall Street Journal*'s front page and the cover of *Inc.* magazine not because of Ben & Jerry's ice cream but because of Ben and Jerry. Having a new ice cream, even one with funnily named flavors, isn't a story, but being "two counter-culture guys . . . trying to build a big business without betraying their principles" is. The *Wall Street Journal's* price for a full-page ad in the same paper in which the Ben & Jerry's story ran was $105, 352. Who can afford to buy that kind of publicity (Coleman and Neville 1993, pp. 268–69)?

Like SKYY in its early years, Racing Strollers, Inc.'s Baby Jogger relies primarily on word-of-mouth and customer usage promotion, taking out only modest ads in *Runner's World* and *Parents.* But the uniqueness of the product has also made it newsworthy, earning stories in publications including the *Wall Street Journal, Washington Post, USA Today,* and the *New York Times* (Thomas 1995, p. 178).

We took full advantage of all the publicity that came our way, from *Newsweek* to the supermarket tabloid *Star.* The latter ran an item trumpeting that "the new Hollywood fad is a brand of vodka, which is supposed to be hangover-free. JACK NICHOLSON loves the stuff and has it delivered to his house." If SKYY hadn't been delivered to his house before, it soon was because we hastily sent him a case—as well as sending copies of the blurb to influential bars and restaurants. We distributed copies of all the articles that were written about us to our distributors and customers—spread the news.

SKYY didn't begin advertising until we were already successful. We didn't use advertising to build the brand; we used it to enhance the brand after free publicity, word of mouth and promotions had already put us on a roll.

When we did begin advertising, we were conservative in some ways, innovative in others. We got

> Advertising in the final analysis should be news. If it is not news, it is worthless.
>
> *—Adolph S. Ochs, publisher and editor*

the most for our money by purchasing ⅓ page vertical ads because an ad the full height of a magazine but just one column wide will be placed next to editorial rather than other ads. We advertised in the kind of magazines we believed had "influential" readers, like *Architectural Digest, Gourmet* and *Wired.* We even ran in *Popular Science* because I believed that the kind of people who read a science magazine would appreciate the science behind SKYY—and tell their friends. We also broke ground by advertising early on in gay publications, such as the *Advocate* and *Out.* As SKYY grew, we moved into full-page advertising.

Events were also key to SKYY's early success. We showed up at every event we heard about—society parties, charity events, symphony galas, fund-raising benefits. We donated cases of SKYY, and walked around pouring people martinis. Who could refuse? We made up some martini glasses with our logo on them and women in evening gowns fought over them.

People who regularly attended such events saw SKYY and became curious about it. They may also have read the brochure we put in their gift bag or noticed that they didn't get the usual morning-after headaches. In any event, they started calling for SKYY at bars and restaurants. They spread the word for us.

We wanted to get SKYY into the hands of influential trendsetters and opinion shapers. We knew that if we got it into their hands once, they'd reach for it themselves the next time. Marketing professionals routinely target their efforts at specific demographics, but I'm talking about something more subtle. Think not only about your end user but also about the people who *influence* your end user. Who are the trendsetters and opinion shapers in your product area? How do you reach them?

This process of influencing various tiers of buyers is sometimes referred to as getting imitators to follow innovators. Business school professor and corporate consultant Robert J. Thomas describes the way highly innovative products gain acceptance. "Innovators, a small proportion of the total potential market, are very eager to buy the new product, and do so. The hope is that imitators will soon follow . . . and boost the volume of sales" (Thomas 1995, p. 8). SKYY's acceptance by innovators, the trendsetters and opinion shapers we initially targeted, fostered a much wider level of consumption

In 1916, Nathan Handwerker promoted his new frankfurters by giving doctors at the nearby Coney Island Hospital free franks—if they ate them at his stand while wearing their white coats and stethoscopes. He hoped the public's respect for doctors would rub off on his product and prove that Nathan's Franks were healthy and high quality (Panati 1987, p. 397).

Who is your target audience? Who do they respect? Who and what influences their buying habits? It's your business to know.

Agencies and consultants use elaborate research to direct and place your ads and promotions. I think we did a pretty good job on our own and you can assemble much of the same kind of information yourself.

If you're trying to determine ideal target regions for launching your product, you can obtain a wealth of information on population through the Census Report and on government Websites (for example: demographia.com). In considering the importation of a new line of scooters, we are using U.S. Department of Transportation information to learn the penetration of motor scooter sales and Census Bureau information to locate regions with the highest percentage of the target audience we have identified (college students and young working adults). This information was free on the Internet.

Before deciding where to place advertising, check the circulation audits of the publications you are considering. Many publications can also provide you with general studies of national buying habits by age, gender, occupation, etc., through MRI surveys.

Trade journals and company Websites can tell you what your competition is doing. Often sales figures, profits and areas of distribution are listed in easy to interpret graphs and charts.

San Francisco's Castro District drinkers understood SKYY's value and embraced its style from the very beginning. And the gay community's influence extends far beyond any neighborhood, affecting popular culture on a variety of significant levels, from fashion and consumer goods to music, movies and every aspect of the entertainment arts. We nurtured this relationship by advertising in gay publications and sponsoring events.

Because I love film and also because the film community wields considerable power to create trends and influence tastes, we cultivated a presence in the independent film world. In addition to sponsoring events at Bay Area film festivals, we sponsored parties at film festivals in Nantucket, Newport, Cannes and Chicago. We went narrow and deep into this market, rather than going wide and shallow into many. We knew that penetrating one or two targeted markets would have a domino effect on other markets.

SKYY has now grown from a locally distributed to an internationally sold brand. Our small four-person office has evolved into SKYY Spirits LLC, a hundred-person plus company with a management team made up of the best people in the business. SKYY is the third best-selling premium vodka in the U.S., the fastest growing brand, and a significant force in the alcohol beverage industry. Not bad for an idea that sprang from a headache.

What I am perhaps most proud of with SKYY is that it's a true example of the enduring power of inventive thinking coupled with an entrepreneurial spirit. As you've read, SKYY's success isn't rocket science; it's based on good common sense and perseverance. As a broke kid with a nifty sweater comb and as a veteran inventor with a "crazy" idea for a new and improved vodka, life has shown me that if you develop a quality product, sell it at a fair price and shoot for an honest profit, you can prosper. The sky really is the limit.

afterword

THE FUNDAMENTAL THINGS STILL APPLY

Inventors don't retire, maybe because inventing doesn't really feel like work for many of us. So what's next for this inventor?

Perhaps you're sipping a SKYY Citrus (try it over ice with a twist) or Vermeer Dutch Chocolate Cream as you read this. Or maybe you're out zipping through traffic on one of our new scooters, wearing your Vee Vest (a lightweight travel vest with lots of zippered inside pockets for maps, passports and cash), pausing when one of my LED stoplights signals you to do so. I hope you'll have started your day with an all-natural, healthy Wagel for breakfast.

Like Sam Farber and his Good Grips utensils, I think I have a few winners left in me. Al and the other folks at my lab can attest to the dozens of projects and patents we have in one stage of development or another.

> When I can no longer create anything, I'll be done for.
>
> *—Coco Chanel, fashion designer and perfumer*

One thing is certain, even in today's rapidly changing New Economy/dot.com/high-tech world, there's still plenty of room for an old-fashioned good idea—one that solves a problem.

I think most inventors at one time or another dream of addressing problems on a vastly different scale than consumer convenience, timesaving or pleasure. I once was privileged to watch a patient regain his vision through a cataract removal procedure with my cryogenic device. Though that process has now been eclipsed by far more high-tech procedures, it remains my proudest achievement. As I write, I am working to organize and fund a free mobile clinic that will perform cataract removal procedures in India, a country I love to visit but also one of the world's poorest. We estimate the cost of each treatment will be no more than twenty U.S. dollars.

Curiosity, stubbornness and some measure of good sense have brought me commercial success, and they've brought me the ability to give back. I don't intend to ever stop asking how and why. I hope I never lose my sense of gratitude as well.

appendix

Creativity, Inventing and Brainstorming

Aha! 10 Ways to Free Your Creative Spirit and Find Your Great Ideas by Jordan E. Ayan (Crown Publishing, 1997).

Biomimicry: Innovation Inspired by Nature by Janine M. Benyus (William Morrow & Co., 1997).

The Complete Idiot's Guide to New Product Development by Edwin E. Bobrow (Alpha Books, 1997). See chapter 4, "Don't Just Sit There Waiting for Godot," for a variety of brainstorming techniques, and chapter 5, "Where Do You Get Those Ideas From, Anyway?"

Five Star Mind: Games and Exercises to Stimulate Your Creativity and Imagination by Tom Wujec (Main Street Books, 1995).

How to Think Like Leonardo Da Vinci: Seven Steps to Genius Everyday by Michael J. Gelb (Dell Publishing, 1998).

Ideas, Inventions and Innovations, available from the Small Business Administration. Consult your local office, call SBA Answer Desk: 800-UASK-SBA (800-827-5722), or log on at www.sbaonline.sba.gov

Peak Learning, rev. ed., by Ronald Gross (Jeremy P. Tarcher, 1999).

A Whack on the Side of the Head: How You Can Be More Creative, by Roger Von Oech, Warner Books, Revised Edition, 1990.

Inventor's Reading List Addresses

Forbes, 800-888-9896, www.forbes.com

Fortune, 800-621-8000, www.fortune.com

New York Times, 800-NYTIMES (698-4637), www.nytimes.com

U.S. Government Printing Office, for publications such as *Survey of Current Business, Economic Report of the President, Business America: The Magazine of International Trade,* and *Official Gazette of United States Patent & Trademark Office of Commissioner of Patents and Trademarks,* issued every Tuesday. Catalog available. 888-293-6498, www.gpo.gov

U.S. News & World Report, 800-436-6520, www.usnews.com

Wall Street Journal, 800-WSJ-8609, www.wsj.com

Prototyping, Fund-raising and Business Plans

Prototyping

CADDCO, concept and CAD drawings. Call 515-547-2867, or log onto www.netins.net/showcase/caddco

California Manufacturing Technology Center, a state agency. Call 800-300-2682 for the office nearest to you.

Castolite, Inc., for mold-making materials and liquid plastics. Catalog available. P.O. Box 391, Woodstock, IL 60098. 815-338-4670.

Dream Merchant magazine, for model maker ads. 310-328-1925.

Edmund Scientific Co., for a variety of parts. Catalog available. 101 E. Gloucester Pike, Barrington, NJ 08007. 800-728-6999, www.edsci.com

Fine Scale Modeler magazine. P.O. Box 1612, Waukesha, WI 53187. 800-533-6644.

Industrial Designers Society of America, 703-759-0100.

McMaster-Carr Supply Co., Los Angeles, CA. A vast assortment of tools and hardware. 562-692-5911, www.mcmaster.com

United States Plastic Corp., Lima, OH. Plastic sheet, tubing, and bar stock, plus adhesives and tools for working with plastics. 800-809-4217, www.usplastic.com

Invention Developers

America Invents, invention firm of highly successful product designer Ken Tarlow. Offers full range of services, including evaluation, design, patent search, writing and drawing services. *Mind to Money* book and tapes available. 21 Golden Hind Passage, Corte Madera, CA 94925, 415-927-0311, www.americainvents.com, email: tarlowdes@aol.com

Arthur D. Little Enterprises, Inc. Contact Manager, New Business Development, 15 Acorn Park, Cambridge, MA 02140, 617-498-6685.

Battelle Development Corporation, 505 King Avenue, Columbus, OH 43201, 614-424-6424, www.battelle.org

Check the National Inventor Fraud Center, www.inventorfraud.com, for red flags to look out for.

Financing Resources

American Venture magazine, a quarterly for entrepreneurs, "angel" investors, venture capitalists, and finance providers. 503-221-9981, www.avce.com

Association of Small Business Development Centers, a co-sponsorship partnership with the SBA, 703-271-8700, www.asbdc-us.org

FinanceHub.com, information and resources online for entrepreneurs and investors, www.financehub.com

Financing the New Venture: A Complete Guide to Raising Capital from Venture Capitalists, Investment Bankers, Private Investors, and Other Sources by Mark H. Long (Adams Media Corporation, 2000).

National Association of Small Business Investment Companies. For a list of financial institutions that deal with small independent businesses and those that work with socially or economically disadvantaged small business owners, write the Association at 666 11th Street NW, Suite 750, Washington, DC 20001. 202-628-5055, www.nasbic.org

Pratt's Guide to Venture Capital Sources, contact Venture Economics, 888-989-8373, www.ventureeconomics.com.

Small Business Administration. Your local Small Business Administration office can be found in the phone book, or call the SBA Answer Desk at 800-UASK-SBA (800-827-5722), www.sba.gov

Small Business Innovation Research makes early stage research and development grants. 800-382-4634, www.sbir.dsu.edu

Business Plans

Business Plan Pro, software by Tim Berry, Palo Alto Software. 800-229-7526, www.businessplanpro.com.

How to Write a Business Plan, 5th ed. by Mike P. McKeever (Nolo Press, 2000).

The Small Business Administration has many business plan resources, including an online training course (www.sba.gov/classroom/bplan914.html); a self-paced, downloadable tutorial called "The Business Plan Road Map to Success (www.sba.gov/starting/businessplan.html); and the publication, *How to Write a Business Plan,* by Linda Pinson and Jerry Jinnett (www.sba.gov/library/pubs/mp-32.doc).

Your First Business Plan: A Simple Question and Answer Format Designed to Help You Write Your Own Plan, 3rd ed. by Joseph A. Covello and Brian J. Hazelgren (Small Business Sourcebooks, 1998).

Documentation, Disclosure and Patenting

Inventor's Journals

Inventor's Journal, Inventions, Patents and Trademarks Co., 888-53-PATENT, www. frompatenttoprofit.com

The Inventor's Notebook by Fred Grissom and David Pressman (Nolo Press, 1996) 800-992-6656, www.nolo.com

Forms

Invention Disclosure format: See Disclosure Document Program instructions at www.uspto.gov/web/offices/pac/disdo.html for requirements and instructions.

Some forms can be downloaded from inventor-oriented Websites, including www.freepatentforms.com

(See pp. 174–76 for some sample forms.)

Patenting

Protecting Your Ideas: The Inventor's Guide to Patents by Joy L. Bryant (Academic Press, 1998).

Trade Secret Home Page, www.execpc.com/~mhallign/

U.S. Patent and Trademark Office, www.uspto.gov, 800-786-9199 or 703-308-4357, for information, patent searches, copies of patents, complete filing and registration information, forms and advice. This should be your first stop and last word. Features Independent Inventor Resources section.

Searches

U.S. Patent and Trademark Depository Libraries: Call for the one nearest to you, 800-PTO-9199.

U.S. Patent and Trademark Office, www.uspto.gov, search current and expired patents.

www.micropatent.com, for searches.

www.patents.ibm.com, for searches.

BASIC CONFIDENTIALITY, OR NONDISCLOSURE, AGREEMENT

Confidentiality Agreement

_____ (your name) _____ agrees to consult with _____ (other party) _____ regarding the unique product _____ (product name or concept) _____.

_____ (other party) _____ agrees to hold in confidence all discussions, trade secrets, recipes, client lists, purveyors, and the like learned during the course of the relationship between _____ (your name) _____ and _____ (other party) _____.

By signing this agreement, each signator affirms that they have the authority to make such agreements on behalf of their companies and that they shall notify all employees, contractors, assignees, agents and designees of the contents of this confidentiality agreement.

____

Name, Date, and Company Title of other party

____

Your Name, Date, and Title

BASIC INVENTOR'S NOTEBOOK PAGE

Name of Invention: ______________________________

Purpose of Invention: ______________________________

____

Description of Invention: ___________________________

____

____

Sketch of Invention: _____________________________

____

____

Unique Features of Invention and Their Advantages Over Existing Technology: ______________________________

____

Actions Taken [discussed with so-and-so, researched materials, made basic model, etc.]: ______________________

Inventor:________________________ Date:__________

This confidential information is witnessed and understood by:

Witness:________________________ Date:__________

Witness:________________________ Date:__________
[second witness optional]

[if notebook pages are not numbered, number them yourself]

MY BINDING ARBITRATION AGREEMENT

ARBITRATION AGREEMENT

In the event that there is a dispute regarding any and all matters relating to our business relationship, it is agreed that the dispute shall be resolved by binding arbitration. It is further agreed that the arbitrator shall not be a lawyer and that the disputants shall not be represented by lawyers, but by themselves only.

We are fully aware of the provision of section 1282.4 of the California Code of Civil Procedure as shown below. (Or of any applicable code if this agreement is signed in any other state.)

[Sec. 1282.4 (Right to Counsel) A party to the arbitration has the right to be represented by an attorney at any proceeding or hearing in arbitration under this title. A waiver of this right may be revoked; but if a party revokes such waiver, the other party is entitled to a reasonable continuance for the purpose of procuring an attorney. Added Stats 1981 ch 460 sec. 2]

I / We hereby, with full comprehension of our rights, agree to waive our rights to be respresented by an attorney at any binding arbitration and shall not ask the court to revoke such waiver.

____________________________________ Date____________

____________________________________ Date____________

Patent Attorneys

Attorneys and Agents Registered to Practice Before the U.S. Patent and Trademark Office, a PTO publication available at libraries.

Obtain attorney recommendations from an inventors association or your local Small Business Administration office.

Manufacturing and Licensing

The Hook Appropriate Technology, helps inventors find marketers and manufacturers, 860-350-2709, www.thehooktek.com

Licensing Executives Society International, www.lesi.org

Manufacturers' Agents National Association (MANA), 877-626-2776.

www.bigbook.com, search site for manufacturers.

www.thomasregister.com, search site for manufacturers.

Also see Wal-Mart, Sharper Image, and Hammacher-Schlemmer references below.

Marketing, Naming and Packaging

Demographic information, check www.demographia.com

800-USA-TRADE, connects you with a trade counselor at the Department of Commerce trained to help American exporters.

Hammacher Schlemmer Search for Invention Contest, www.hammacherschlemmer.com

How I Made $1,000,000 in Mail Order, rev. and updated ed. by E. Joseph Cossman (Fireside, 1993).

Infomercial Marketing Report, call 310-826-8810 for a sample issue of this monthly publication.

Marketing Without Advertising by M. Phillips and S. Rasberry (Nolo Press, 1997).

Radical Marketing by Sam Hill and Glenn Rifkin (Harper Business, 1999).

The Sharper Image "is always interested in innovative, unique products, and . . . encourages you to submit your ideas for consideration." For instructions, see www.sharperimage.com, or write The Sharper Image, 650 Davis Street, San Francisco, CA 94111.

Target Marketing magazine, all about direct marketing. www.targetonline.com

Under the Radar: Talking to Today's Cynical Consumer by Jonathan Bond and Richard Kirshenbaum (John Wiley and Sons, 1998).

Wal-Mart Innovation Network (WIN), for inventions with zero to six-month sales histories, 417-836-5671, www.win12.com

Packaging

New York Design Studio, 516-569-8888.

T2Design, 310-656-9922, www.members.aol.com/t2design/

Trademarks

NameStormers computer software, 214-350-6214.

Patent and Trademark Office, see their Website for search information, or visit one of the Patent and Trademark Depository Libraries.

Thomson & Thomson, www.thomson-thomson.com, for professional trademark searches including TrademarkScan, online and on CD-ROM. Services Support: 888-477-3447.

Trademark: How to Name Your Business and Product by Kate McGrath and Stephen Elias (Nolo Press, 1992).

Inventor Resources, Associations and Assistance

From Patent to Profit workshops based on the book of the same name. A variety of other resources are available, 888-53-Patent, www.frompatenttoprofit.com

International Federation of Inventors' Associations (IFIA), www.invention-ifia.ch

National Congress of Inventor Organizations, P.O. Box 93669, Los Angeles, CA 90093, 213-878-6952, www.inventionconvention.com

United Inventors Association of the USA, P.O. Box 23447, Rochester, NY 14692, 716-359-9310, www.uiausa.org

Websites

The Internet Invention Store, www.inventing.com, showcases inventions and new products, lists "products in need of investors, manufacturers, or licensing," and provides lists of inventor service providers.

InventNET, www.inventnet.com, The Inventor's Network.

National Inventor Fraud Center, www.inventorfraud.com, "Helping inventors through education." Includes BAD guys and GOOD guys lists.

Patent Café, www.PatentCafe.com, features Ask the Experts Message Board, Live Expert chats, and downloadable invention assessment forms.

Newsgroups

alt.inventors, post questions and join in discussions.

misc.int.property, the legal side of patents, trademarks and copyrights.

patent-news, send subscribe, unsubscribe, help and information requests to administrative address: Majordomo@world.std.com

Magazines

Dream Merchant, 2309 Torrance Blvd., Suite 104, Torrance, CA 90501, 310-328-1925.

Entrepreneur magazine, 949-261-2325, www.entrepreneurmagazine.com

Inventor's Digest, 800-838-0808. Also online at www.inventorsdigest.com

Trade Shows and Conventions

Invention Convention, 323-878-6925, www.inventionconvention.com

National Inventors' Expo, maintained by the U.S. Patent and Trademark Office.

Tradeshow Central, www.tscentral.com

bibliography

There are many resources aimed at helping inventors. To ensure that my experiences and advice might be useful and instructive, I compared notes with and sometimes cited the following:

The Baby Jogger Website, babyjogger.com

Bobrow, Edwin E. *The Complete Idiot's Guide to New Product Development.* New York: Alpha Books, 1997.

Coca-Cola Website, www.cocacola.com

Coleman, Bob, and Deborah Neville. *The Great American Idea Book.* New York: W. W. Norton & Company, 1993.

Debelak, Don. *Entrepreneur Magazine: Bringing Your Product to Market.* New York: John Wiley & Sons, 1997.

DeMatteis, Bob, with Mark Antonucci. *From Patent to Profit: Secrets & Strategies for Success.* Inventions, Patents & Trademarks Co., 1997.

Elias, Stephen. *Patent, Copyright & Trademark.* 2nd ed. Edited by Lisa Goldoftas. Berkeley, CA: Nolo Press, 1997.

Gelb, Michael J. *How to Think Like Leonardo Da Vinci: Seven Steps to Genius Every Day.* New York: Dell Publishing, 1998.

Gleick, James. "Patently Absurd." *New York Times Magazine,* 12 March 2000.

Goldratt, Eliyahu M. *The Goal: A Process of Ongoing Improvement.* 2nd rev. ed. Great Barrington, MA: North River Press, 1992.

Griffith, Richard, and Arthur Mayer. *The Movies.* New York: Simon and Schuster, 1970.

Hirasuna, Delphine. "Sorry, No Beige." Apple Media Arts, 1998, online (www.apple.com/creative/collateral/ama/index/html)

Kawasaki, Guy, with Michele Moreno. *Rules for Revolutionaries: The Capitalist Manifesto for Creating and Marketing New Products and Services.* New York: HarperBusiness, 1999.

Macdonald, Anne L. *Feminine Ingenuity: Women and Invention in America.* New York: Ballantine Books, 1992.

MIT inventor's Website, www.mit.edu

"Needle Epidemic—Actions, Reactions." *San Francisco Chronicle,* 14 April 1998.

Newhouse, Elizabeth L., ed. *Inventors and Discoverers Changing Our World.* Washington, DC: National Geographic Society, 1988.

Nolo's Legal Encyclopedia, www.nolo.com

OXO Good Grips Website, www.oxo.com

Panati, Charles. *Panati's Extraordinary Origins of Everyday Things.* New York: Harper & Row, 1987.

Perrier Website, www.perrier.com

Pressman, David. *Patent It Yourself.* 6th ed. Berkeley, CA: Nolo Press, 1997.

Reynolds, Pat. "SKYY Gets It Right." *Packaging World,* December 1994.

Science Friday, "Technology Patents." National Public Radio, 24 March 2000.

Shekerjian, Denise. *Uncommon Genius: How Great Ideas Are Born.* New York: Penguin Books, 1990.

Thomas, Robert J. *New Product Success Stories: Lessons from Leading Innovators.* New York: John Wiley & Sons, 1995.

Winship, Michael. *Television.* New York: Random House, 1988.

acknowledgments

My heartfelt gratitude and admiration go to Yvette Bozzini for her intelligent, astute, and creative research and advice and for providing the perfect backup whenever one of my best stories hit a bald patch.

Thanks also to Al Kolvites, Tom Hunter, and Melissa Lilly for reading and commenting on parts of the manuscript; to Sam Farber and Phil Baechler for the straight scoop on their inspiring stories; and to Jeff Campbell for careful editing and suggestions.

index

advertising, 148–150, 153–154, 161–163
advice
  faulty, 96
Apple, 142–143
arbitration, 121
audit, 122

Baby Jogger, 22, 32–33, 52, 161
Baechler, Phil, 32, 52
bankruptcy, 123
bottles, 141–148
brainstorming, 26, 139, 169
brochures, 158–159, 162
business plan, 55
  resources for, 54, 172
buyers, 5, 39, 163

cataracts, 168
choice, 14
cinema. *see also* films
  distributors for, 18
  multiplex, 18
  Quad, 9–15, 18–20, 27, 44, 104, 128, 137
circulation, 164
classification number, 87
clay, 45
collaboration, 145
competition, 158
confidentiality, 83–85, 114–115, 130, 174
congeners, 126–128, 131–132, 152
connections, 21, 30, 32
consignment, 131
contests, 12
contingency, 95
copyright, 80
creativity, 169
  fostering of, 21–22
curiosity, 1, 20, 22
customer(s)
  fill needs of, 57, 65
  find out about, 57, 59–62
  identify with, 41, 43, 56, 59
  information from, 57–58, 66
  product to, 155–156
  as sales rep, 153

David Susskind Show, 6
dentists, 37–38
design, 56–57, 61
  basic, 43
  collaboration with, 145
  industrial, 143
  intuitive, 145
  problems and, 118
  simple, 48–49, 73
  students of, 149

studying, 148–149
universal, 23–24
details, 146–148
development, 67–69
D-Fuzz-It, 1, 5, 7, 13, 21, 27, 50, 95, 104, 108, 110, 131, 132, 145, 146, 148, 152
directories, 115
disclosure statement, 84–85
distribution, 38–40, 51–52, 54, 109, 118
consumer and, 155–156
help in, 152–153
international, 153–154
marketing and, 157–165
distributors, 18, 59, 110, 132, 134
documentation, 82, 173
due diligence, 114–116

events, 162
exit strategy, 120
experts, 111
use of, 28

fads, 90
familiarity, 22–23
Farber, Sam, 23–24, 32–33
feedback, 141–142
film(s), 31. *see also* cinema
attendance at, 17–18
festivals of, 164–165
independent, 18
financing, 4, 49–52, 54–56
resources for, 53, 171–172
focus groups, 57, 59

giveaways, 60–61, 159–160
Good Grips, OXO, 23–24, 30, 47, 66
graphic designer(s), 4, 12
guarantees, 131

habit, 19
headaches, 128–129, 132–133
honesty, 124

IBM, 87
idea(s), 2, 12
feasibility of, 44
implementation of, 35
protection of, 82–85
reworking of, 66–67
imitation, 13, 144–145
improvements, 20–21, 101, 112, 149
income, 103
percentage of, 106
India, 61
innovation, 14, 19, 27, 142–143, 163
instinct, 145
Internet, 7, 29. *see also* Websites
information from, 87, 163–164
interstate commerce, 136
interviews, 65, 111
invention(s), 2, 169–170
assistance companies for, 50
associations for, 177
conception of, 84
customary for, 48
disclosure statement for, 84–85
fear of losing, 59
improvement on, 90
improvements as, 20–21
journals for, 82–83, 173
modification of, 67
proving of, 43–46
reinventions and, 33–36
review process of, 56
techniques for, 21–22

inventor
as employee, 105–106
marketing by, 157–158
inventor service organizations, 50, 96
investor(s), 50–51
journals, 83–84, 173

Kanbar Target, 37–40, 59, 66, 82, 107, 109
knowledge, 25–27
knowledge
power of, 30
labels, 146–148

lawyers, 2, 75, 85–86, 87–88, 93–97, 119–120, 123
learning, 25–26
LED, 30–31, 167
legal encyclopedia, online, 115
library, 86–87
licensing, 54, 74, 90, 92, 105, 107, 175
agreement for, 121–123
disputes under, 120–121
due diligence of, 114–116
exclusive/nonexclusive, 119–120, 121
exit strategy of, 120
market scope of, 122
pitch for, 116–119, 124
royalties with, 119–121, 123–124
special, 108
upfront fee for, 120, 123
licensee
track record of, 115
litigator, 94
loyalty, 135

management, 54
manufacturing, 3, 35, 47, 77, 105, 175
control of, 106
established corporation and, 107, 109, 117
hiring outside, 109–113
production cost for, 109
quality of, 111
single product, 106–108
market research, 10, 42, 56–57, 57–62, 66, 68, 89–90, 117–118
low cost, 58
pricing and, 150–151
marketing, 31, 35, 38–39, 48, 56–57, 69, 108 109, 175–176
advertising and, 161–163
consumer and, 155–156
established corporation and, 130
events and, 162–163
feedback from, 141–142
infomercial and, 156
inventor's involvement in, 157–158
niche, 63
packaging and, 141–149
promotional sales and, 156
promotions and, 60–61, 159–160
publicity and, 160–161
quality of product and, 131
target audience for, 163
marks
registered/unregistered, 137
media. *see* publicity
medical instruments, 24
megaplex. *see* cinema
metal
case, 46
sheet, 45

Micropatent, 87
model(s), 2, 12, 96. *see also* prototypes
  inexpensive, 43–44, 47–49
  maker of, 50
  materials for, 45–46
  modifications to, 45
  production, 44–45
  working, 117
mold(s), 3–4, 74
  inexpensive, 47–48
  owning, 109
  working, 123
motivation, 10–11

name(s), 128–129, 135–140
  distinctive, 12
needle. *see* safety needles
nondisclosure agreement, 83–85, 92, 111, 114–115, 174

observation, 16–17, 18–19, 20, 22, 24, 61–62
opportunities, 17, 24
orders, 5
OSHA (Occupational Safety and Health Administration), 74

packaging, 4, 134, 176
  design of, 142–146
  generic, 141
  printing and, 146–147
  second-rate, 148
Patent and Trademark Office (PTO), 44, 77–78, 81, 82, 92, 129
  applications and, 88
  classification number and, 87
  Depository Library of, 86–87
  Disclosure Document Program for, 85
  filing of trademark with, 136–137, 146
  infringement and, 90
  licensing and, 116
  maintenance fees to, 92
  online search through, 87
  trade dress registration with, 144
patent(s), 74, 105
  agents for, 84, 87
  application for, 3, 88–93, 97
  attorney, 2, 75, 85–86, 87–88, 93–97, 119–120, 123, 173
  broad claims of, 97
  criteria for, 77
  definition of, 78
  design, 79–80
  distinction of, 79
  documentation for, 82–83
  first, 2–3
  first to invent for, 82
  foreign countries and, 88
  history of, 76–77
  improvements on, 76
  infringement of, 85, 89–90, 122
  pending, 81, 92, 106, 111, 114, 117
  plant, 80
  prior sales and, 89
  prior to process of, 82–85
  profits and, 89
  protection with, 38
  provisional application for, 91–92
  sale of, 119–120
  search for, 85–87, 94, 96, 173
  strategy for, 90–91
  utility, 79, 97

Perrier, 136, 138–140
physics, 31
pitch, 116–119, 124, 155
plans
  reviewing of, 34–35
plastic, 45
press release, 160
pricing
  mark-up and, 149
  strategy for, 150
printing, 146–147
problem(s)
  determining of, 26
  observed, 24
  solving, 9–15, 22–23, 26, 33, 36, 69, 168
product
  consumer and, 155–156
  definition of, 122
  development of, 26, 149
  distinctive, 142, 144
  expanding lines of, 113
  improvements of, 112, 117, 149
  naming of, 128–129, 135–140
  pitching of, 116–119, 124, 155
  pricing of, 149–151
  quality of, 131
  research on, 65
  second-rate, 148
  testing of, 60, 150–151
  twist for, 129
progress
  review of, 8
promotions, 60–61, 156, 159–160
prototype(s), 3, 37–41, 86, 170–171.
  see also models
  development of, 43–44
  patents and, 90
  resources for, 46
  testing of, 42, 56, 67
publicity, 12, 131–134, 152, 160–161, 177

questions, 2, 17, 71

reading, 28–29
recordkeeping, 82–83
Rex Products, Inc., 62
right of first refusal, 122
ROLLOcane, 40–43, 66, 107, 109, 155–156
royalties, 114–121, 123–124

safety needles
  medical, 71–75, 82, 97
sale(s), 3, 51, 103
  estimate, 117
  international, 6
  packaging influence on, 144
  promoting, 152–153
sales reps, 5, 153
samples, 4, 123, 153
samurai, 146
scenarios, 10
science, 25. *see also* physics
scooters, 62–64, 66–67, 167
simplicity, 48–49, 73
SKYY Bounce Game, 21
SKYY Vodka, 7, 17, 51, 60, 64, 68–69, 80, 101, 104, 109, 111, 112, 125–134, 137–138, 140 ,141–148, 152–167
Small Business Administration (SBA), 52
Snapple, 154
soft drinks, 154–155
solutions
  inventing, 30

Starbucks, 64–65
steamer, 65–66
students, 47, 62, 149
success
  5 steps to, 7–8, 56–57

Tangoes, 102–104, 107, 111–113, 117
Tangram, 99–102
television
  rise of, 9–11, 17–18
testing, 42, 56, 60, 67, 150–151
theatres, 9–15
trade dress, 144–145, 148
trade secret, 81, 92, 106, 115, 129–130
trade shows, 47, 62, 115, 178
trademark(s), 80, 146, 176
  filing for, 136
  search of, 137
trendsetters, 63, 162–163
typeface, 5

United Inventors Associations of the USA, 93
Universal Design, 23–24

value, 150, 159
venture capitalist, 52–54
Vermeer, 67, 140, 151, 153, 164, 167
vision, 31
vodka. *see* SKYY Vodka

Websites, 177
wood, 45